IMAGES
of America

THE BIG E
EASTERN STATES EXPOSITION

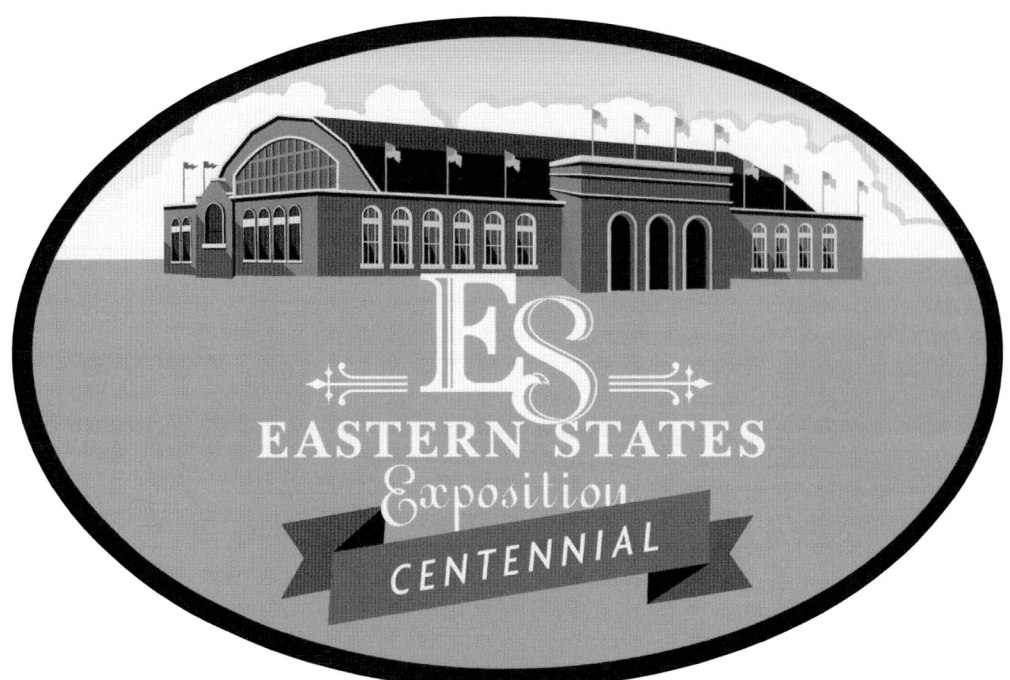

This sticker commemorating the centennial of the Eastern States Exposition features a new logo for the organization and an illustration of the Coliseum arena, the first structure built on the grounds in 1916. Both were created by graphic designer Deborah Walsh of TSM Design in Springfield, Massachusetts. (Courtesy of Deborah Walsh.)

ON THE COVER: The scene of livestock judging, horse shows, rodeos, graduations, hockey games, circuses, concerts, and other events, the Eastern States Exposition's Coliseum has been a local landmark for the past 100 years. (Courtesy of Eastern States Exposition.)

Images of America

The Big E
Eastern States Exposition

David Cecchi

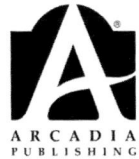

Copyright © 2016 by David Cecchi
ISBN 978-1-4671-1716-6

Published by Arcadia Publishing
Charleston, South Carolina

Printed in the United States of America

Library of Congress Control Number: 2016931077

For all general information, please contact Arcadia Publishing:
Telephone 843-853-2070
Fax 843-853-0044
E-mail sales@arcadiapublishing.com
For customer service and orders:
Toll-Free 1-888-313-2665

Visit us on the Internet at www.arcadiapublishing.com

Dedicated to Marilyn Curry, who I am honored to call my friend.

CONTENTS

Acknowledgments		6
Introduction		7
1.	Off to a Great Start	9
2.	Water, Wind, and War	53
3.	Show Window of the East	85
4.	Becoming The Big E	105
Bibliography		127

ACKNOWLEDGMENTS

What is The Big E? September, searchlights, state buildings, and 17 days. Funland, 4-H, and FFA. Poultry, produce, and Maine baked potatoes. Crafts, corn dogs, and country singers. Beads, butter sculptures, Baby Beef, and the Better Living Center. Grand champion geese, the Grange, and Grinderama. Hallamore Clydesdales, Hurricane Hell Drivers, and helicopter rides. Millie's pierogis, marching bands, and Hitler's armor-plated Mercedes. Log houses, Little Irvy, largest horses, and Agawam Lions Club chicken dinners. Stake Night, superstars, and Storrowton Village. Tractors, trustees, and turkey legs. Gate 9, governors, and the Giant Slide—just to start.

There is so much to see and do (and eat!). More than just a fair—bigger than six state fairs—the Eastern States Exposition became known as "The Big E" for good reason. The exposition does a truly outstanding job of simultaneously evolving while maintaining its finest traditions.

What follows is my tribute to "the fair." It was impossible to picture every aspect, activity, and innovation of this amazing institution, but I do hope that you enjoy what does appear on the following pages.

Sincere thanks go to Noreen Tassinari, the exposition's director of marketing. When I approached Noreen about this project, she graciously and enthusiastically offered full access to exposition archives—a treasure trove of images, artifacts, and information.

I would also like to thank Eastern States Exposition (ESE) president Gene Cassidy, ESE communications manager Catherine Pappas, ESE marketing assistant Sue Gallup, and ESE senior writer and social media coordinator Anne-Alise Pietruska for their assistance with this project. Dennis Picard, director of the exposition's Storrowton Village Museum, came through in the eleventh hour with some elusive information. Thank you!

Bill Fearn very generously shared photographs, materials, and memories of his involvement with the 4-H and Baby Beef program at the fair. Betty Blascak Denehy, Todd Collier, the late Everett Hodge, Ellen Janik, Lenny Lapinksi, Richard Lind, Peggy Lis-Barone, Karen Miller Grant, Nancy Neilson, Linda Noftall, Alice Smith, Patty Souder, and Deborah Walsh provided assistance or shared photographs, materials, and information useful in the compilation of this book. Thank you.

While every attempt has been made to present accurate information, responsibility for any erroneous material is entirely my own. Unless otherwise noted, all photographs are courtesy of the Eastern States Exposition.

Until September,
David Cecchi
Agawam, Massachusetts
January 2016

Introduction

The 1964 edition of the Eastern States Exposition opened to cloudy skies. Running from September 19 to 27, the fair, for the second consecutive year, was granted official "International Fair" status by the US Department of Commerce, one of only two fairs in the country so designated that year. Making appearances in 1964 were Arthur Godfrey and Goldie, his Palomino stallion; the Royal Inniskilling Fusiliers, Royal Ulster Rifles, and Royal Irish Fusiliers; the exposition's World Championship Rodeo; the Royal Canadian Mounted Police; the Eighth Air Force Band; and David Cecchi.

I remember nothing about that visit, my first (I was just an infant), nor the next few—but I have attended the exposition every year since. I graduated from Agawam High School among the wood shavings of the Coliseum—one of the last five Agawam classes to do so. I took part in the exposition's first Agawam Day on September 14, 1989, as a member of the Agawam School Committee. I introduced my sons Joe and Bailey—who both later entered youth giant squash and pumpkin contests—to the fair. I have roamed the grounds as a spectator and, for more than three decades, as an exhibitor—along with my brothers Bobby and Michael, annually entering produce from the family farm, E. Cecchi Farms in Feeding Hills, Massachusetts, in the Native Produce Display (and often garnering top honors). I was awfully excited to be invited to the exposition's Community Breakfasts kicking off the countdown to their 100th anniversary. In case you have not figured it out, I love the exposition.

This is not a comprehensive, detailed history by any means. I am not sure such a thing could even be possible—there is just too much to the Eastern States Exposition. Innovations and firsts—the exposition and those involved with it have influenced so much over the past century and, really, continue to do so. At any rate, I highly recommend Frances Gagnon's wonderful *Eastern States Exposition 1916–1996, An Illustrated History at 75 Years.* I also expect that *The Republican's* newest title in its Heritage Series, commemorating the Eastern States Exposition's centennial and written by *Republican* executive editor Wayne Phaneuf, will also be a must-have for any exposition aficionado.

It is, of course, images that will drive this narrative of the exposition, and credit must be given to those whose work appears on the following pages. Professional photographers Xenophon A. Beake, Beake-Huntington Inc.; Lloyd White Bell; Warren Boyer; Ray D'Addario; Vincent S. D'Addario; Ralph R. Doubleday; Fairchild Aerial Survey Inc.; Fay Foto Service Inc.; Robert F. Hildebrand; Hodge & Sampson; Mario Sarno; Huntington & Napolitan Inc.; Paul Krause; Clarence A. Rice, Rice & Watkins; H.A. Strohmeyer Jr.; and Woodhead Photo Co. Inc., are all represented here, along with several amateur and unidentified photographers. We are fortunate that the exposition has been so well documented, in images and in print.

While the Eastern States Exposition marks its centennial in 2016, the genesis of the "improvement" movement from which it sprang dates back to the first decades of the 20th century. In 1911, John A. Scheuerle, a Hartford, Vermont, pastor, and others, inspired by the Boston 1915 and Better Brattleboro movements, sought to improve local rural conditions. The resulting Hartford, Forward! movement accomplished much to make life in that small Vermont village of just over

4,000 residents "economically prosperous, physically healthy, intellectually satisfying, civically sound, and socially delightful."

With an eye toward expanding the success of the Hartford, Forward! movement to a larger scale, the Bennington County Vermont Improvement Association was organized the following year, with Scheuerle serving as secretary, a role that today would be known as director. Initially focused on road improvement, the scope of the organization soon expanded to include the betterment of agriculture, education, social services, and civic concerns.

It was at this time that Horace A. Moses, who was tackling similar issues in the village of Woronoco, Massachusetts, the site of his Woronoco Paper Company mill, took notice of Scheuerle's efforts in the Green Mountain State. A correspondence soon followed, the result of which was the formation of the Hampden County Improvement League in 1913, with Scheuerle serving as secretary/director. The objectives of the league, as outlined in its Articles of Incorporation, were:

> To acquire and disseminate information regarding modern agricultural activities; to increase the productivity of farms; to bring about cooperation in producing and marketing farm products. To arouse and stimulate sentiment for road improvement; to bring about efficient road administration, and systematic work for the maintenance of roads within the county of Hampden. To encourage the schools to introduce such subjects and pursue such activities as will fit the boys and girls to become more efficient in the activities in which they are to engage. To encourage and promote the federation and cooperation of all community forces in the adoption of long term policies. For work in economic, educational, recreational and civic betterment; for the encouragement and rendering more profitable the pursuit of agriculture and, generally to foster, encourage, and promote all things in the communities of Hampden County which tend to increase the productivity of the soil, or to advance or conserve the education, civic, and moral welfare of the communities.

While the league set about this noble work, New England agriculture continued its decline that began in the middle of the 19th century. From 1860 to 1910, a 42-percent reduction in cultivated acres, combined with less productive farms and a population increase of 110 percent, resulted in the importation of nearly 80 percent of the region's food supplies. The net effect of this was a general increase in the cost of living and a loss of over $7 billion dollars in capital (in 2016 dollars) from the region.

Several prominent local businessmen involved in the league's efforts took particular note of the deplorable state of agriculture throughout the Northeast. One of these men was Joshua Loring "J.L." Brooks, founder of the Brooks Bank Note and Lithographing Company and president of the Springfield Board of Trade, who felt strongly that "unless New England's farmers are successful, her industries will suffer."

Brooks's efforts to improve the state of the region's agriculture, provide a method of cooperative purchasing, and foster greater cooperation between agriculture and industry were part of what was known as the Eastern States movement. Brooks conceived of a massive regional agricultural and industrial exhibition that would spread the gospel of this movement. When Brooks shared his idea with Theodore N. Vail, president of the American Telephone & Telegraph Company, Vail declared, "This plan is not only sound, but it appears vital to the future of New England."

It was with this endorsement that Brooks's exhibition took shape as the Eastern States Agricultural and Industrial Exposition. Chartered under Massachusetts law in 1914, organizers hoped to host an inaugural event in West Springfield, Massachusetts, during the fall of 1916 that would permanently establish the exposition as a prominent and important agricultural event.

That event, the National Dairy Show, did just that, and set the course of the exposition for the next century.

One
OFF TO A GREAT START

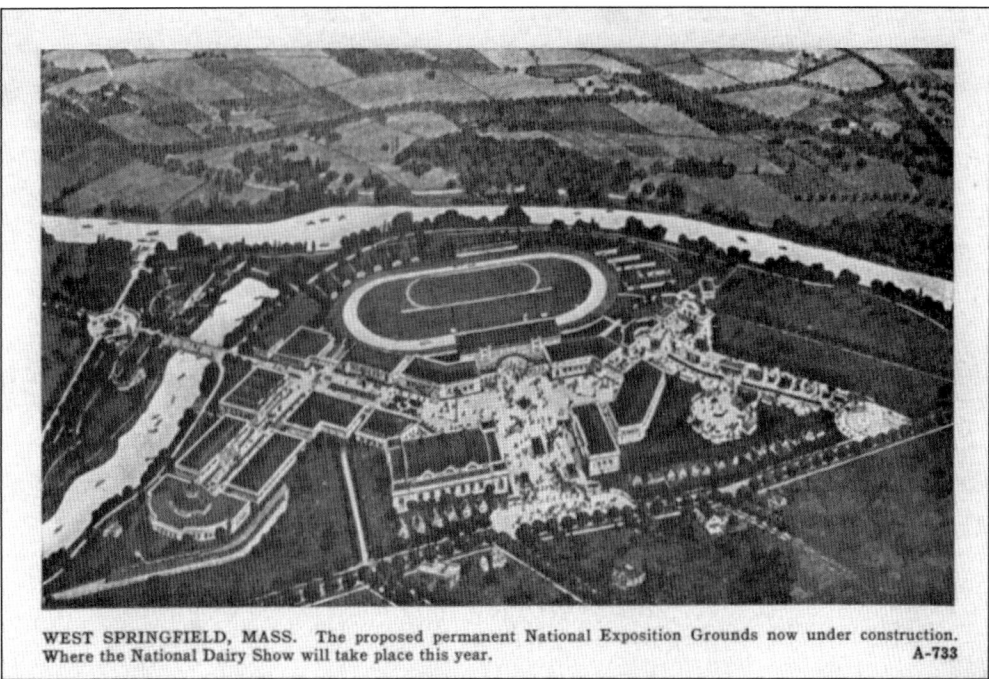

WEST SPRINGFIELD, MASS. The proposed permanent National Exposition Grounds now under construction. Where the National Dairy Show will take place this year.   A-733

This promotional postcard predates the 1916 opening of the exposition and features an artist's rendition of what the fully built-out fairgrounds might look like. This early design was heavily influenced by the City Beautiful movement launched by architect Daniel H. Burnham with the construction of the "White City" at the 1893 World's Columbian Exposition in Chicago. Note the dirigible floating above the open fields of neighboring Agawam, Massachusetts. (Author's collection.)

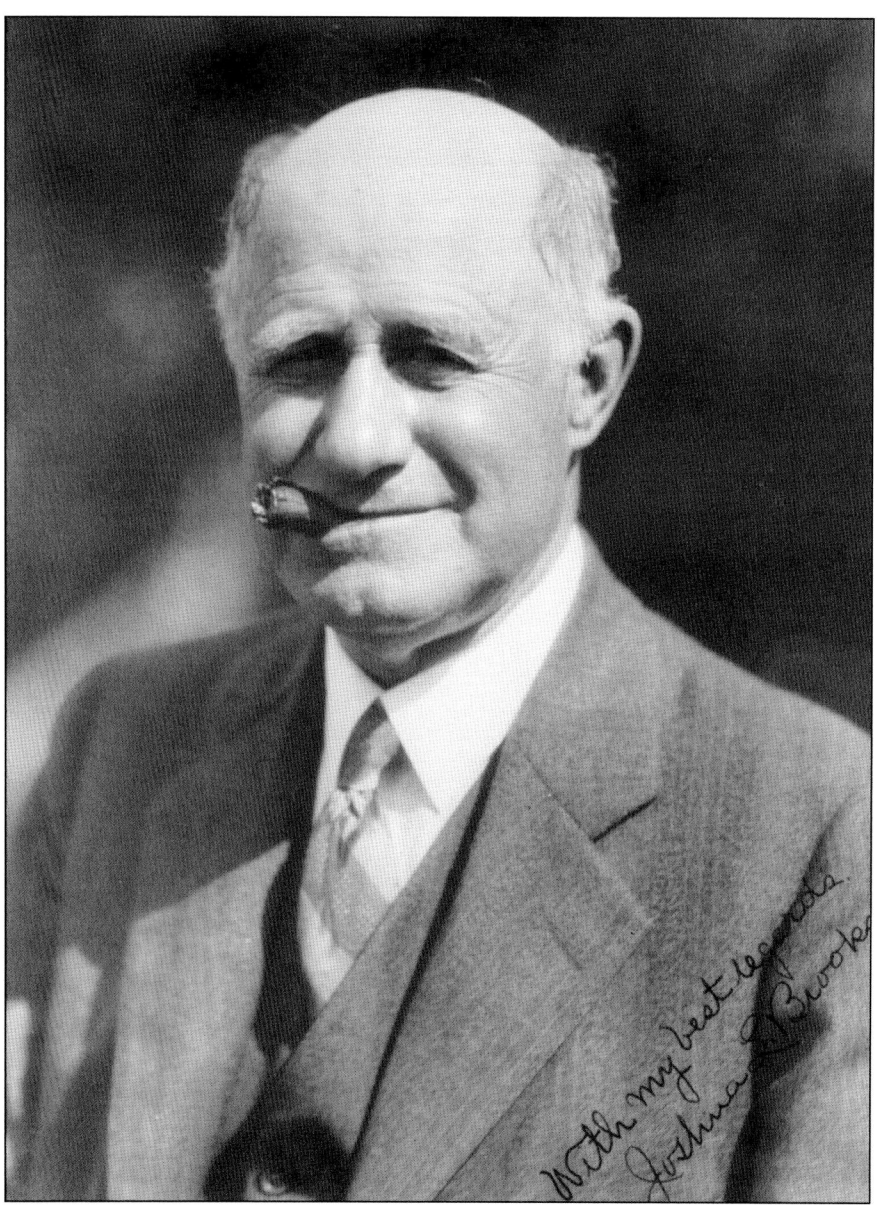

Joshua "J.L." Brooks, founder of the Eastern States Agricultural and Industrial Exposition, was its president from 1916 to 1942. Brooks founded the J.L. Brooks Bank Note and Lithographing Company in Boston in 1889 and moved the firm to Springfield in 1898. By 1913, the Brooks Bank Note and Lithographing Company was being promoted as "the best equipped lithograph house in the country." That same year, Brooks was elected president of the Springfield Board of Trade; it was in his position as business leader and owner of a 300-acre Wilbraham dairy and horse farm that he became acutely aware of the codependence of agriculture and industry, which ultimately led to the founding of the exposition. On display in this photograph is Brooks's ever-present cigar; as unbelievable as it is today, doctors had advised him to take up smoking as a way to relieve stress. Longtime Lewiston (Maine) *Evening Journal* editor Arthur G. Staples once said of Brooks, "He is a bundle of energy—never still—and an unlighted, half-smoked, well-chewed cigar butt is his constant badge of activity."

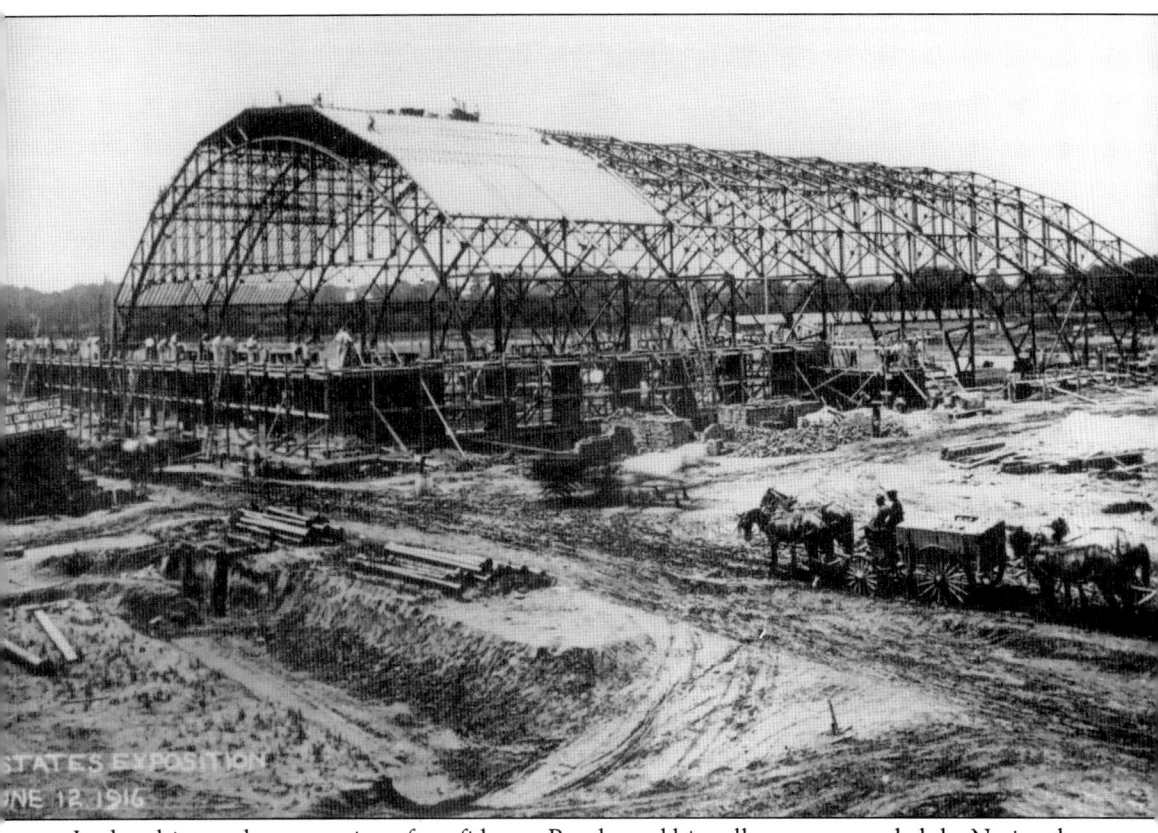

In the ultimate demonstration of confidence, Brooks and his colleagues persuaded the National Dairy Association to hold its 1916 national show in Massachusetts, despite the fact that no facilities yet existed at the West Springfield fairgrounds. With less than 10 months until opening day, Springfield's Fred T. Ley & Company was chosen to construct the exposition's massive Coliseum, larger by more than 1,000 square feet than New York City's Madison Square Garden. The first 200-foot steel arch was erected on May 18, 1916, with the 10th and final arch put in place 10 days later, setting a record and garnering national attention. Fred T. Ley & Company was justifiably proud of this achievement and highlighted the feat in its advertising. Fred T. Ley & Company later became one of the nation's largest construction companies and in 1929 constructed the Art Deco masterpiece Chrysler Building in New York City. The Coliseum, or "Colly" as it is affectionately referred to by exposition staff, is a local landmark.

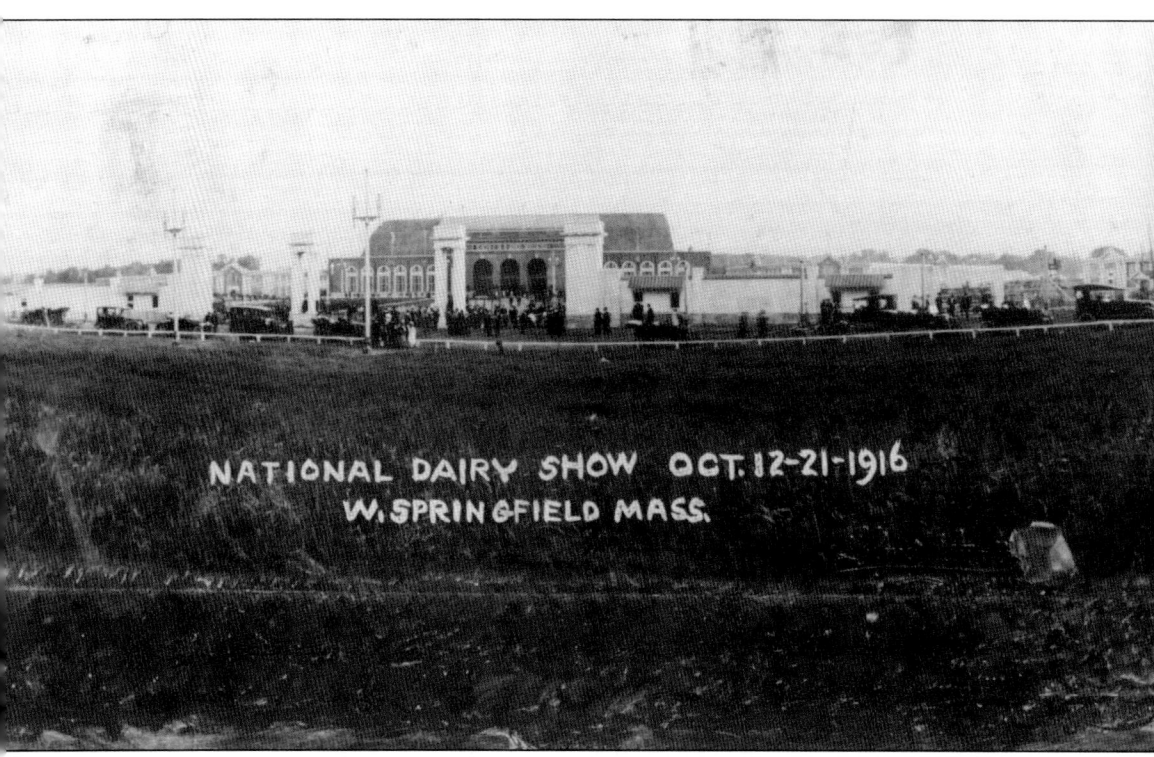

Construction of the Coliseum was finished in time for the opening of the National Dairy Show on October 12, 1916. With a seating capacity of 5,000, no structure like it existed. Other structures completed for the show included Machinery Hall (now C Barn), the cattle barn and horse building (later replaced with the Young Building and the Stroh/Farmarama Building), and the Women's Building (now the New England Center). The show was lauded in special "National Dairy Show Number" and "Lithogravure" sections of the *Springfield Republican*. Among the animals assembled for the show were 24 purebred Jerseys from the famous Ed C. Lasater herd. At 1,600 head, it was at the time the world's largest herd, situated on his 360,000-acre Falfurrias, Texas, ranch, along with 900 other dairy cows and 86,000 beef cattle. Also making an appearance was Royal Figgis Fox o' Dreamwold, the 1916 New York State Fair grand champion bull, from Thomas W. Lawson's Dreamwold estate in Egypt, Massachusetts. (Author's collection.)

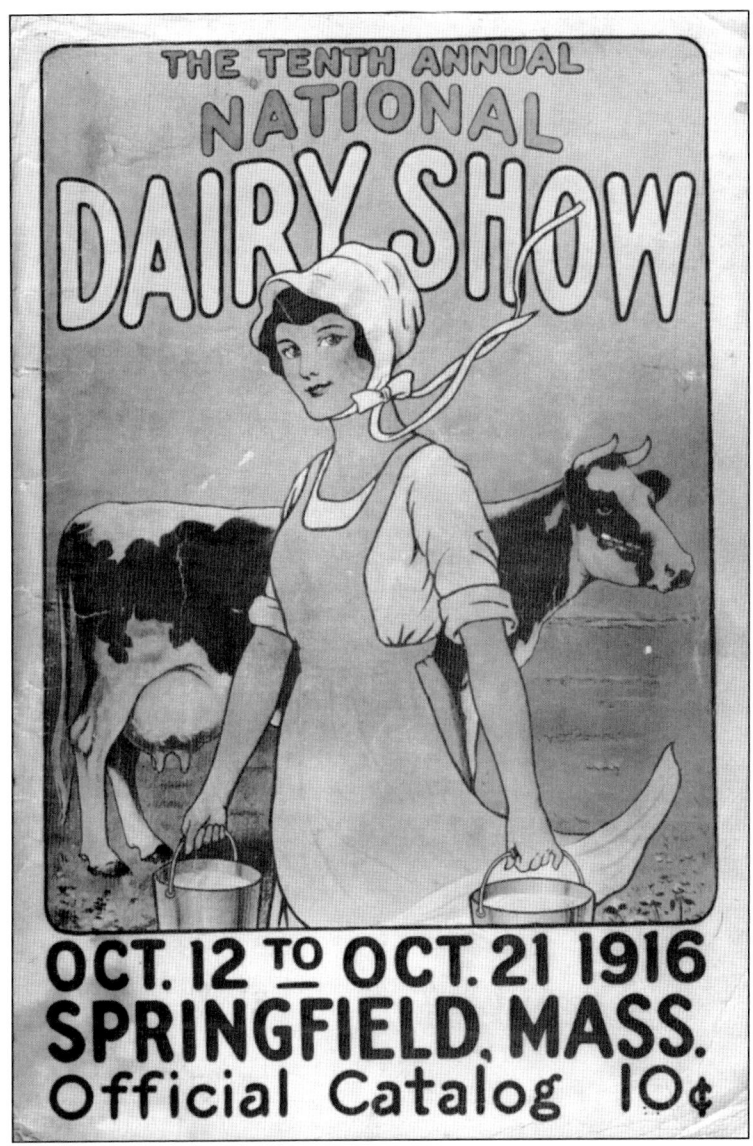

When the 10th annual National Dairy Show opened in West Springfield on October 12, 1916, it was the first time the show had been held east of Chicago. Conventions, conferences, and meetings scheduled during the show included, on October 16, the International Milk Dealers' Association Convention, in the Mahogany Room at the Municipal Auditorium (Symphony Hall); Men in Charge of Cow Testing Associations in the Coliseum Convention Hall; and the Official Dairy Instructors' Association at Technical High School's Assembly Room. The Conference of County Agents met on October 17; the next day, the Ayrshire Breeders' Association Banquet and the US Department of Agriculture Conference of Committee from Breed Associations and Representatives from the Dairy Division took place. The National Dairy Show Association held its annual meeting at the Hotel Kimball on October 19. The New England Federation for Rural Progress also met that day in the Coliseum. October 20 saw the Conference of State Jersey Cattle Organizations at the Hotel Kimball. The National Dairy Herdsmen's Association Luncheon took place on October 21, the closing day of the show. (Author's collection.)

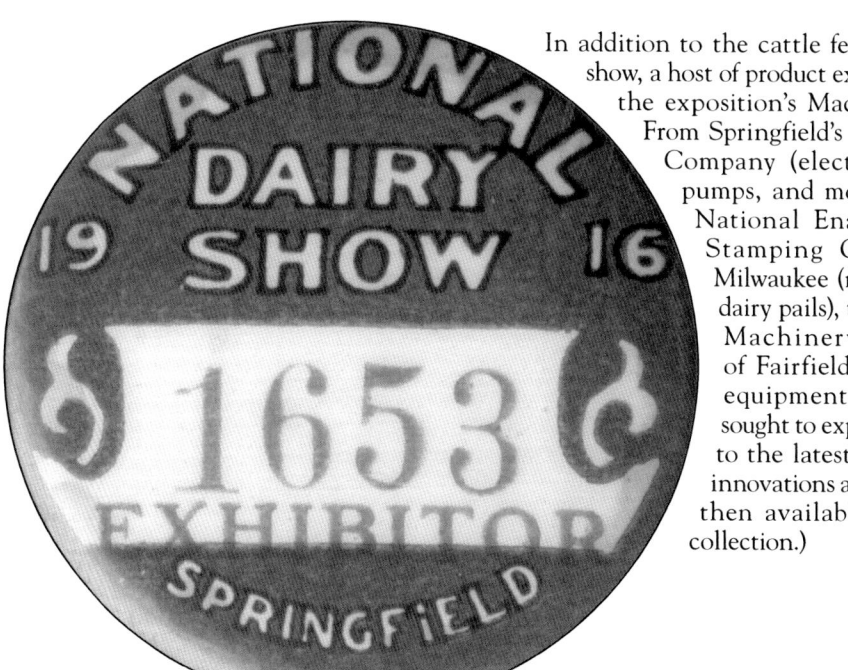

In addition to the cattle featured at the show, a host of product exhibitors filled the exposition's Machinery Hall. From Springfield's Bacon-Taplin Company (electric lighting, pumps, and motors), to the National Enameling and Stamping Company of Milwaukee (milk cans and dairy pails), to the Louden Machinery Company of Fairfield, Iowa (barn equipment), exhibitors sought to expose dairymen to the latest and greatest innovations and equipment then available. (Author's collection.)

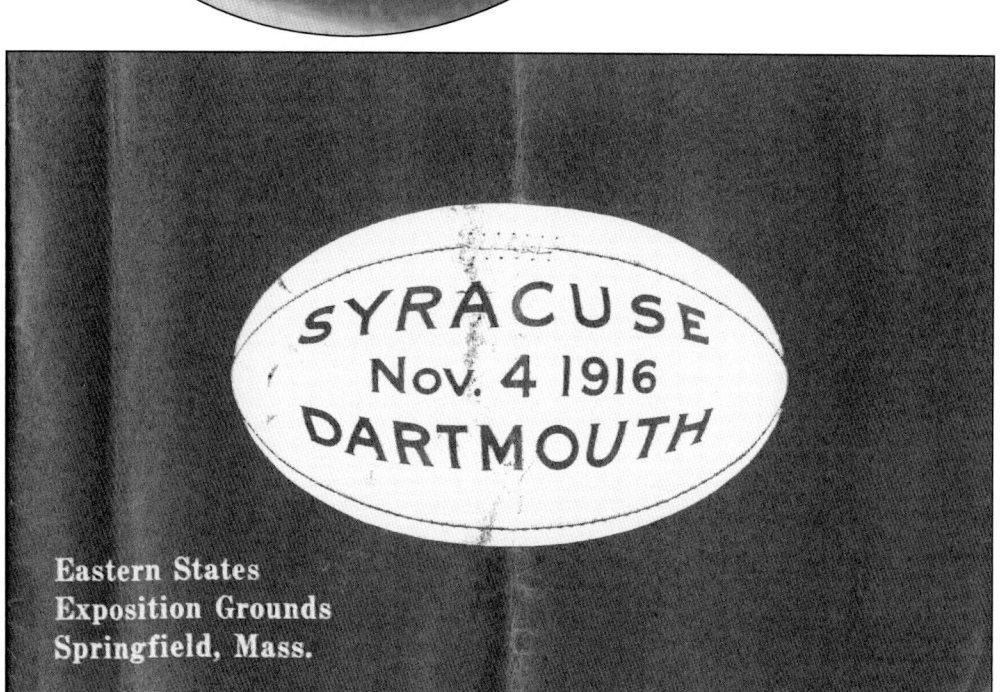

An autumn fair/exposition was always envisioned as just one component of more expansive offerings by the exposition. A 1916 football game between Dartmouth College and Syracuse University was witnessed by 12,000 spectators, including the legendary ballet dancer Vaslav Nijinsky. He and the Ballets Russes were chauffeured to the game in a fleet of Cadillac touring cars; later that evening, they performed at Springfield's Court Square Theater. In the game, Dartmouth prevailed 15-10. (Author's collection.)

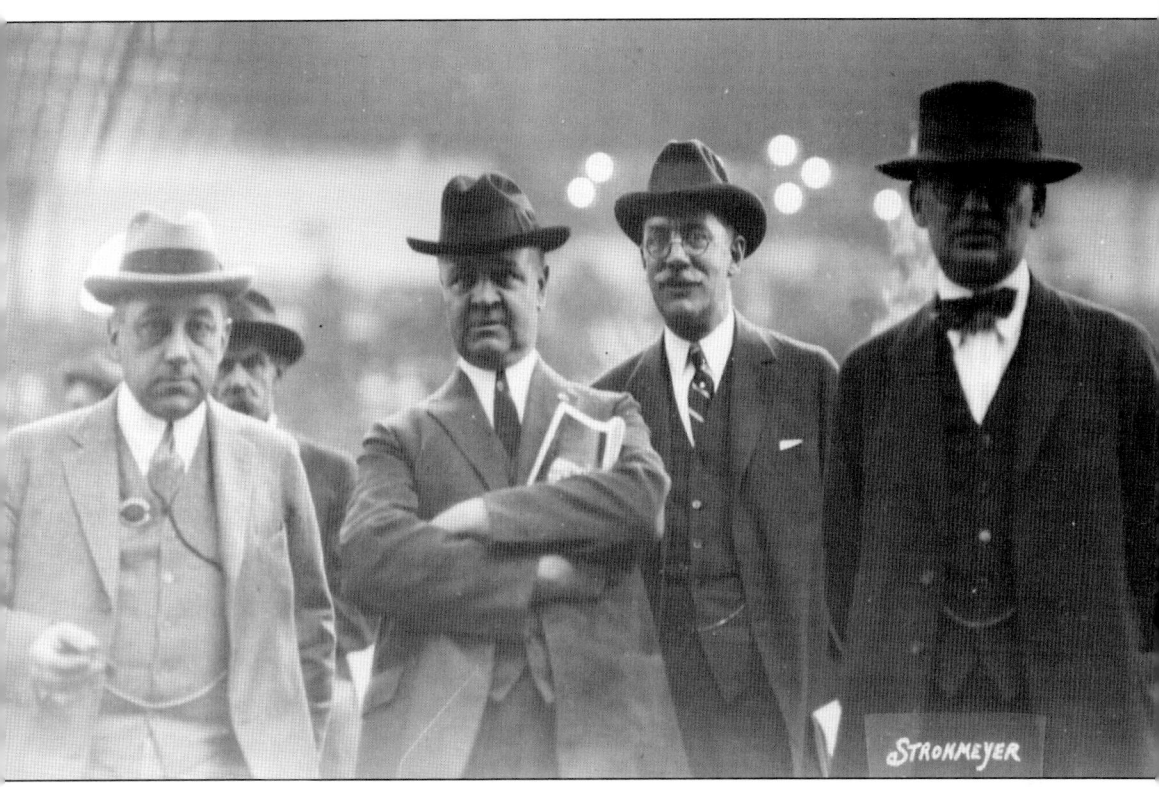

John C. Simpson (far right) became the first general manager of the Eastern States Exposition on January 1, 1917. Previously, he had been manager of the Minnesota State Fair. The *Springfield Republican* reported that Simpson was said "to have had greater experience in the conduct of agricultural fairs than any other man in the country" and was considered "the livest wire in the live Northwest." Simpson served as general manager until 1922. With Simpson are, from left to right, Robert Scoville, Hampden E. Tener, and unidentified. Scoville was a member of the board of directors of the National Dairy Association and was elected president of the American Guernsey Cattle Club in 1918, a position he held for several years. Scoville also served as a trustee of the Eastern States Exposition from 1917 through 1934. Tener began his career as an assistant to Andrew Carnegie and was later an organizer of the Fidelity Trust Company. He also owned Walnut Grove Farm in Washingtonville, New York, where he bred champion shorthorns. He was an exposition trustee for more than two decades.

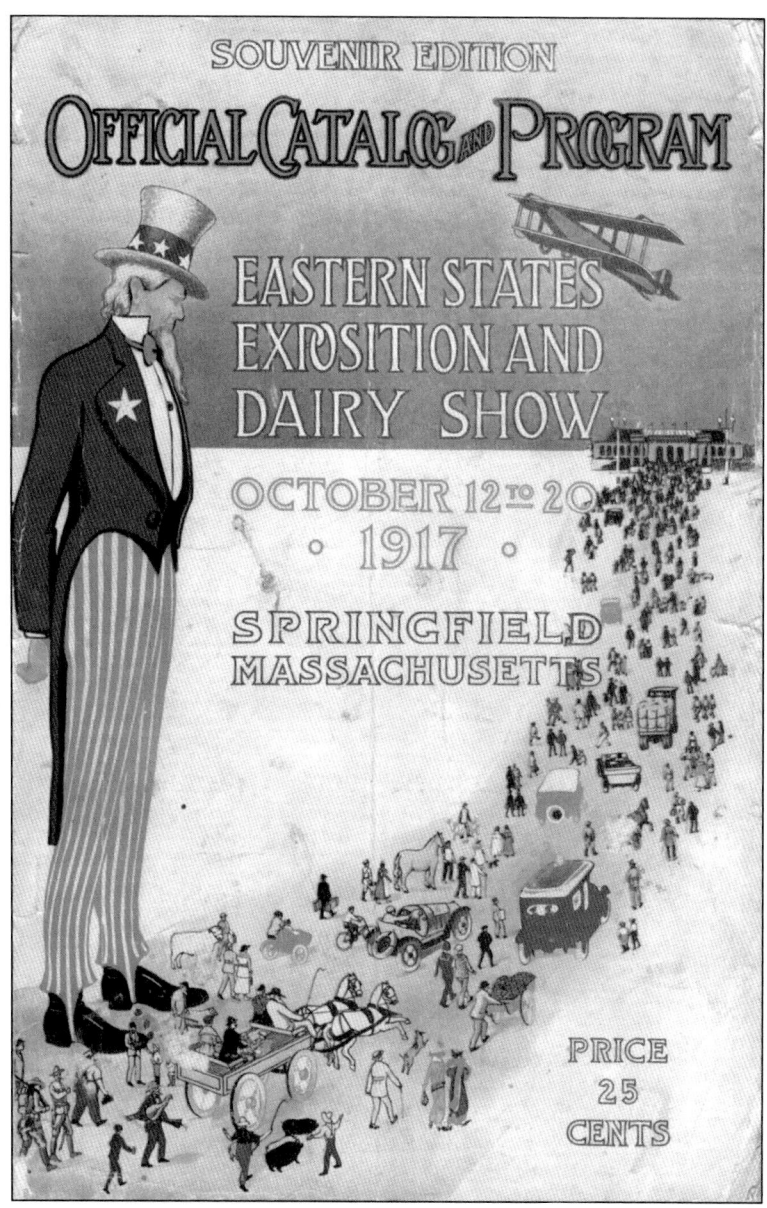

Buoyed by the success of the previous year's event, the 1917 Eastern States Exposition and Dairy Show ran from October 12 to 20. That year's program, of decidedly patriotic design, explained that the exposition was "designed to illustrate the advances, not of one state, not of New England . . . but of all the states of this . . . most populous section of the Union. We have organized an Exposition which typifies the new spirit of this region—the spirit which brought into the foreground "Bill Jones," the new type of Eastern States farmer and dairyman." The program also noted that the first Industrial Exposition and Export Conference "yet attempted" was held on the exposition grounds from June 23 to 30, 1917. Part of the event was a "program of practical talks on the general subject of After the War—What?' that had and will have large permanent value." The first National Vegetable Show of the Vegetable Growers' Association of America was held at the exposition in 1917. That year, the group celebrated its 10th anniversary and held its annual convention in Springfield during Exposition Week. (Author's collection.)

Vermont was an early and enthusiastic supporter of the exposition. In this issue of *The Vermonter* magazine, Charles R. Cummings expressed astonishment at "the enormous size and permanent construction of the buildings . . . at the broad avenues of tarred stone—at the general expansiveness of things." He said of Vermont's exhibit in Machinery Hall, "When you can get near enough through the crowd . . . what do you see: a few familiar people handing out maple sugar as if their lives depended on it; a great fireplace effect in the rear of the expansive space—only it isn't a fireplace—it's a couple of great intertwined horns or cornucopias about a great yellow beehive and discharging upon the floor—a great outpouring of vegetables and fruit twenty-five feet across, in blending color. Were you ever so proud of your little state in your life?" (Both, author's collection.)

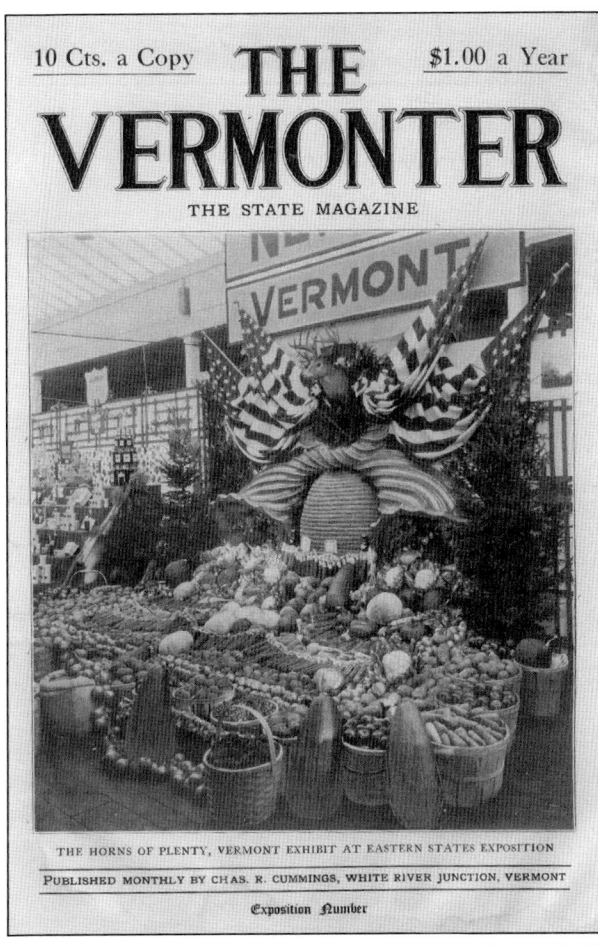

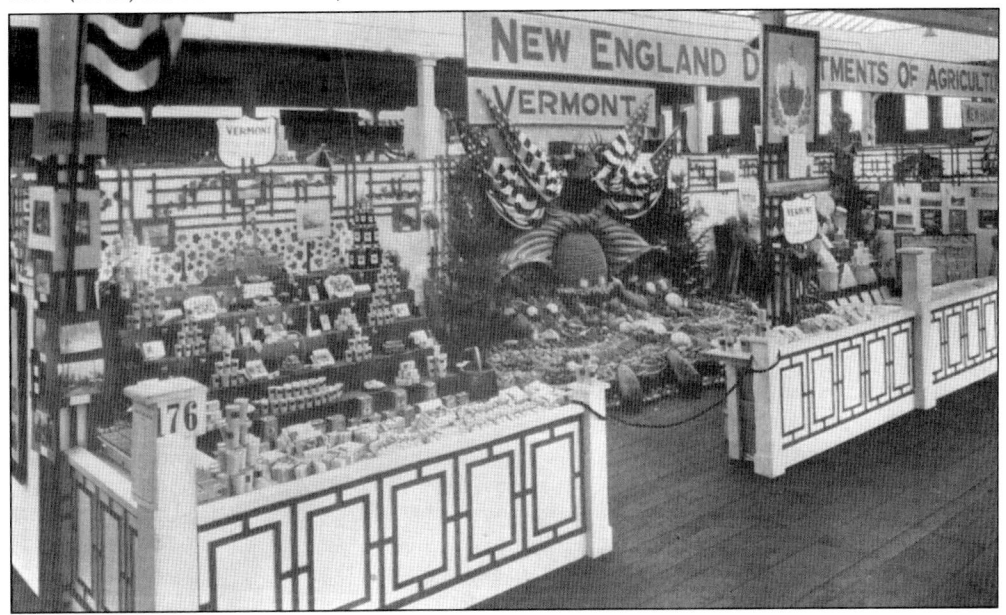

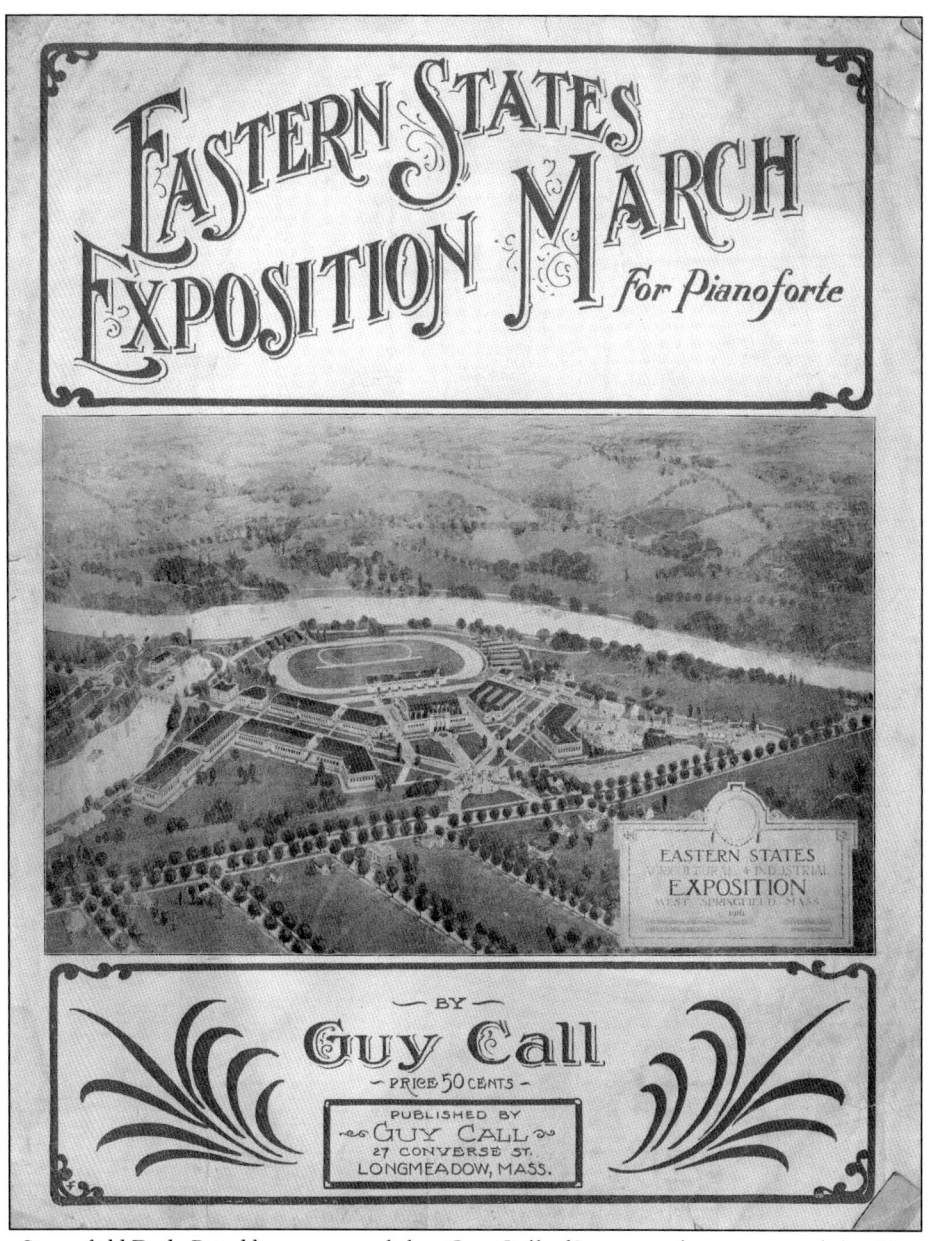

The *Springfield Daily Republican* reported that Guy Call of Longmeadow composed the "Eastern States Exposition March" after being "inspired by reports of the magnitude of the National Dairy Show" and "stirred by the extraordinary achievements of the Eastern States Exposition workers." At the time, Call was employed by the Springfield department store Meekins, Packard & Wheat, as well as being a "talented musician" who played with the Court Square Theater orchestra. The paper also stated that Call was "in no way interested in cattle breeding." Call would later go on to become principal bass player with the Springfield Symphony Orchestra. His march was performed on the opening day of the National Dairy Show by Short's Concert Band. The cover of the sheet music depicts the exposition grounds as envisioned by landscape architect Albert Davis Taylor of Cleveland, Ohio, only a portion of which had been built at that time and which would ultimately not be fully completed as designed. (Author's collection.)

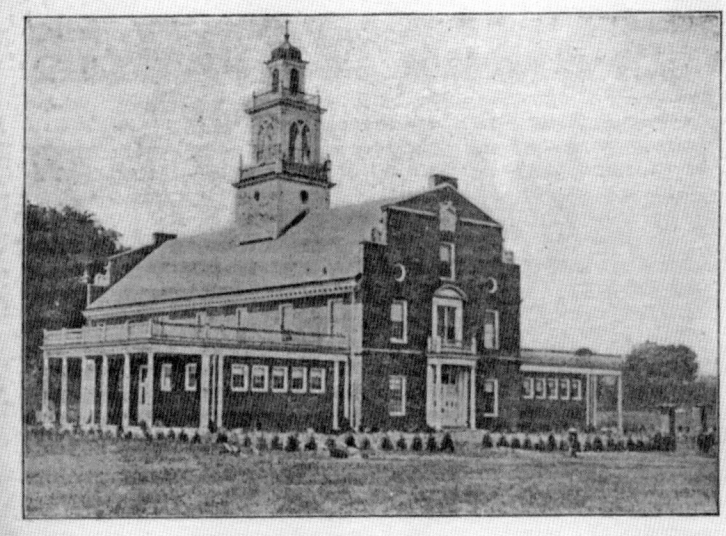

No fair was held in 1918, as its buildings were used for storage by the government during World War I. The 1919 exposition, declared *The Eastern States Magazine*, a bimonthly publication available by subscription for $1 per year, would be "a greater show than ever before." Volume I, No. I, from September 20, 1919, explained that the magazine was created "to further the activities and to voice the aims and accomplishments of The Eastern States Agricultural and Industrial Exposition, Inc., The Eastern States Agricultural and Industrial League, The Eastern States Farmers' Exchange, and all other organizations and movements which may affiliate with them to advance the agricultural and industrial interests of New England and other Eastern States." The inaugural issue featured the new Massachusetts State Building on the cover. Within were articles by league president Horace A. Moses, US Secretary of the Interior Franklin K. Lane, O.H. Benson of the US Department of Agriculture, and Kenyon L. Butterfield, president of the Massachusetts Agricultural College (now the University of Massachusetts Amherst). (Author's collection.)

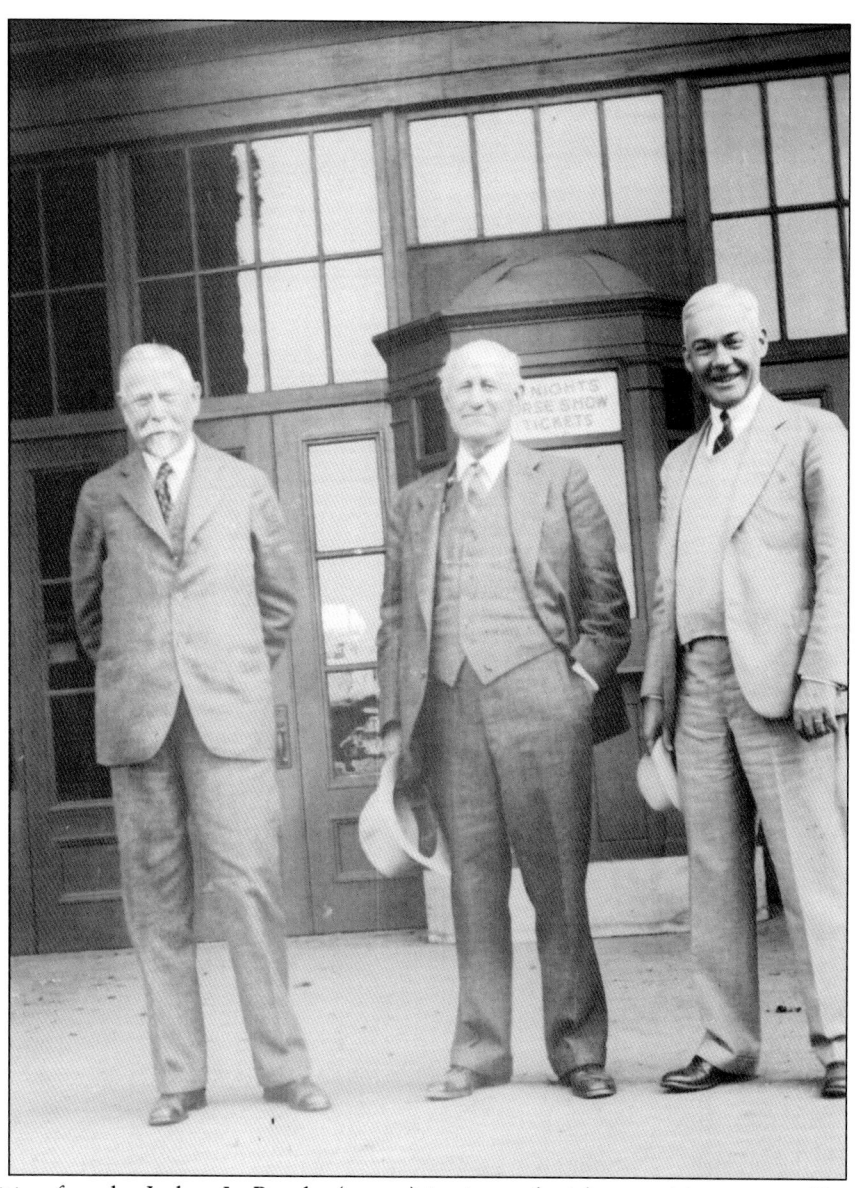

Exposition founder Joshua L. Brooks (center) is pictured with exposition trustee Horace A. Moses (left) and Charles A. Nash. In 1916, Brooks and Moses were awarded the Pynchon Medal and inducted into the Order of William Pynchon (the region's highest recognition of service to the community) by the Publicity Club of Springfield (now the Advertising Club of Western Massachusetts). Moses, "in recognition of your service to the community in having organized and brought to its present state of efficiency and influence the Hampden County Improvement League, to your zeal and devotion to the upbuilding of the agricultural resources of New England . . . believing that the future prosperity of not only Hampden County, but all New England, will be . . . influenced by the work you have done," and Brooks, for "service to Springfield and the community in having conceived and brought to fruition, what we believe to be one of the most important undertakings [the Eastern States Exposition] ever attempted for the material development of Springfield and the community of which it is the natural center." Nash was assistant manager of the fair from 1917 to 1923 and general manager from 1923 to 1951.

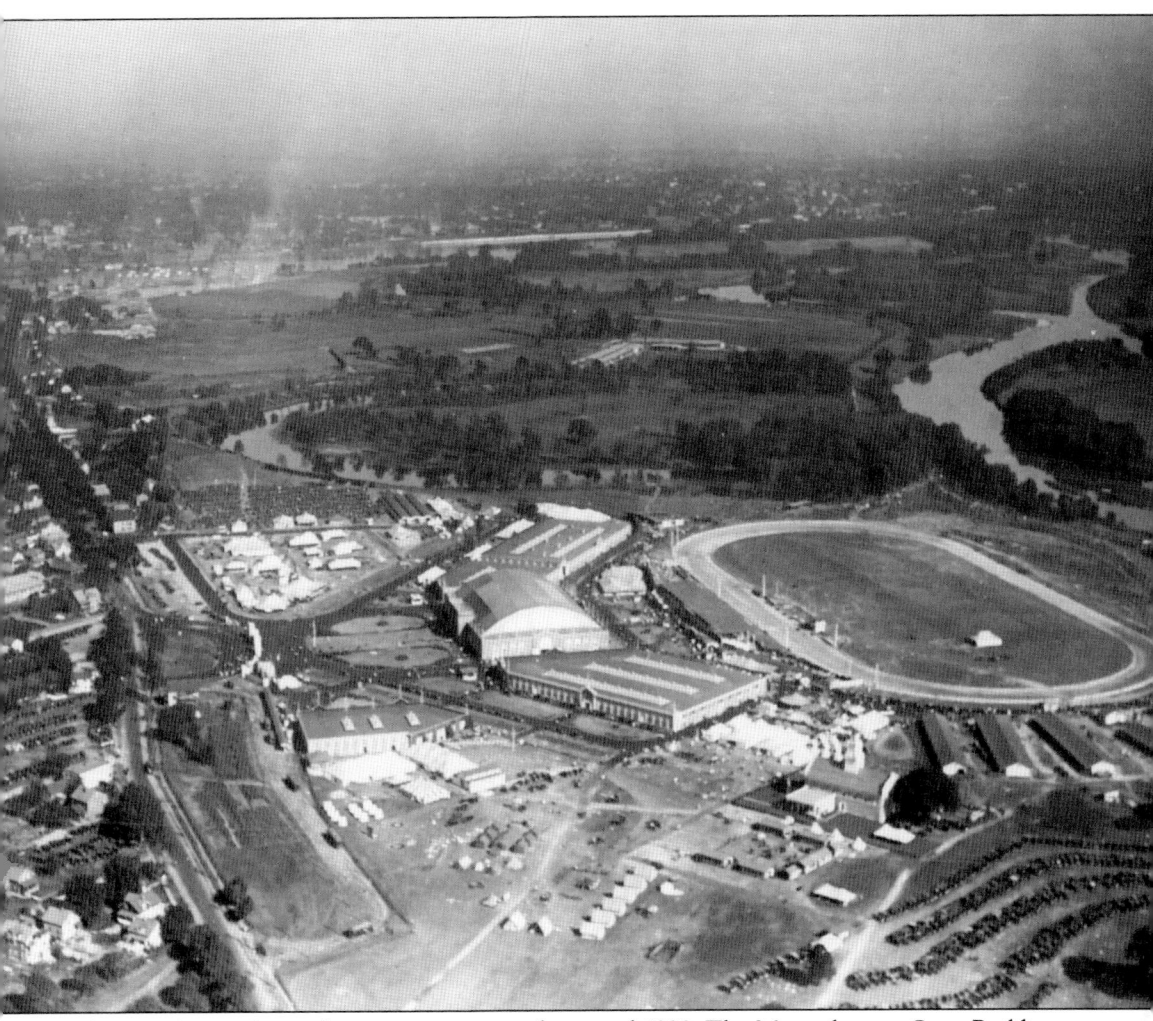

This aerial view shows the exposition grounds around 1920. The Massachusetts State Building at lower right stands alone on what would become the Avenue of States. Among those attending its September 16, 1919, dedication was Massachusetts lieutenant governor Channing H. Cox and a delegation from the local chamber of commerce. At mid-left, just below the main entrance to the fair, is the trolley station. Also visible at right is the racetrack, which was used for both automobile and horse racing and, sometimes, according to exposition president Gene Cassidy, automobiles and horses racing against each other. The first racing event of the 1919 fair, held on September 20, featured "Farmer" Bill Endicott and Louis Disbrow, both veterans of the first three Indianapolis 500 races, competing in a three-mile race. Disbrow won by inches. He had as a passenger the *Springfield Republican* sports writer; Endicott carried the sports writer from the *Springfield Union*.

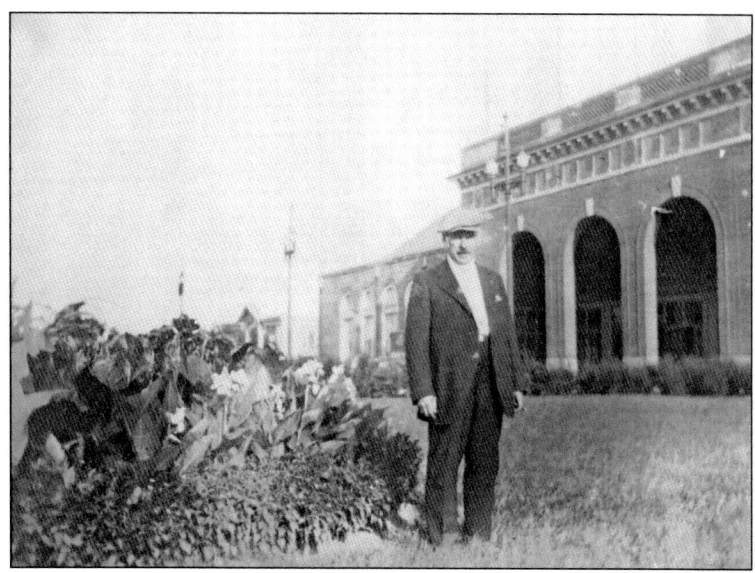

In the early years of the exposition, there were lush gardens located in front of the Coliseum, where the current Court of Honor Stage is now located. The man in the photograph is George Stockwell, stable superintendent. Later, a small grove of evergreens was planted at this site, which has since been removed.

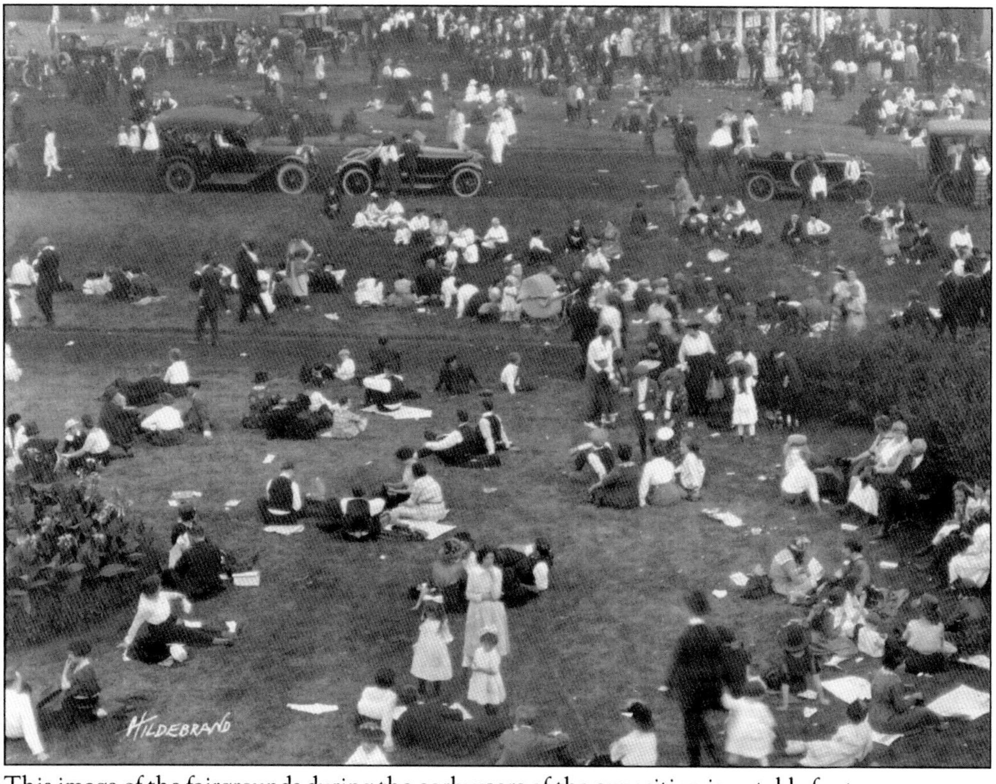

This image of the fairgrounds during the early years of the exposition is notable for two reasons—the fact that automobiles apparently were allowed onto the grounds among the crowds and the amount of litter strewn about. Were there no trash barrels anywhere?

Camp Vail was one of the five divisions of the exposition's Boys' and Girls' Department and represented boys' and girls' club work "to attain a richer and more satisfactory rural life through better farm and home practices." Participating youngsters gave daily demonstrations and lived in a semi-military camp setting during the fair. (Author's collection.)

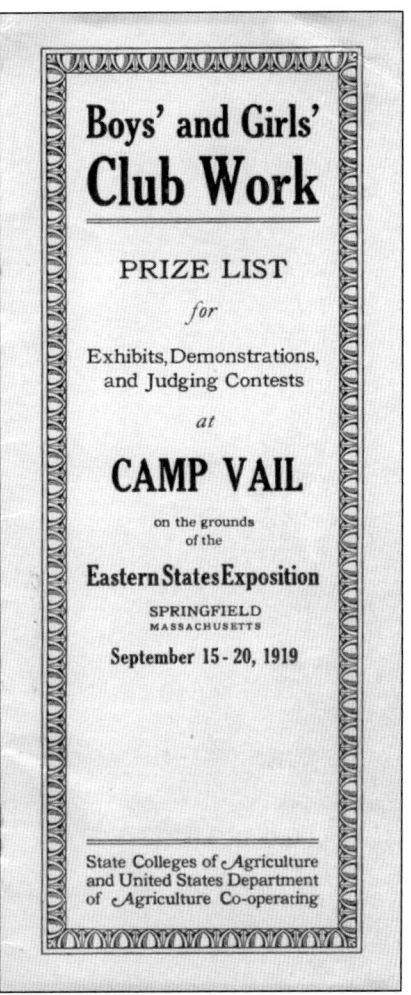

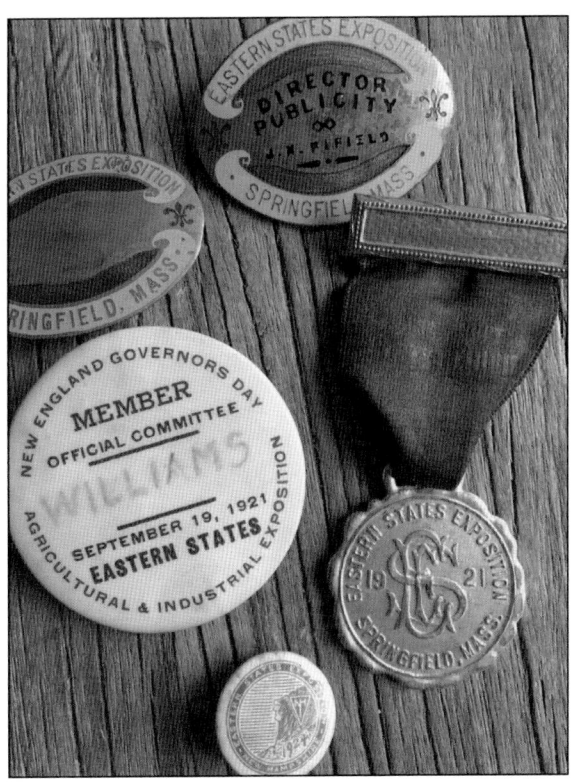

A selection of exposition pins and badges from the 1920s includes that of James H. Fifield (top right), advertising and publicity director at the exposition from 1925 until 1941. At left is a Governors Day official committee member pin from 1921—the first year of the event. At bottom is a commemorative pin from the 1924 New Hampshire exhibit in the Industrial Arts Building.

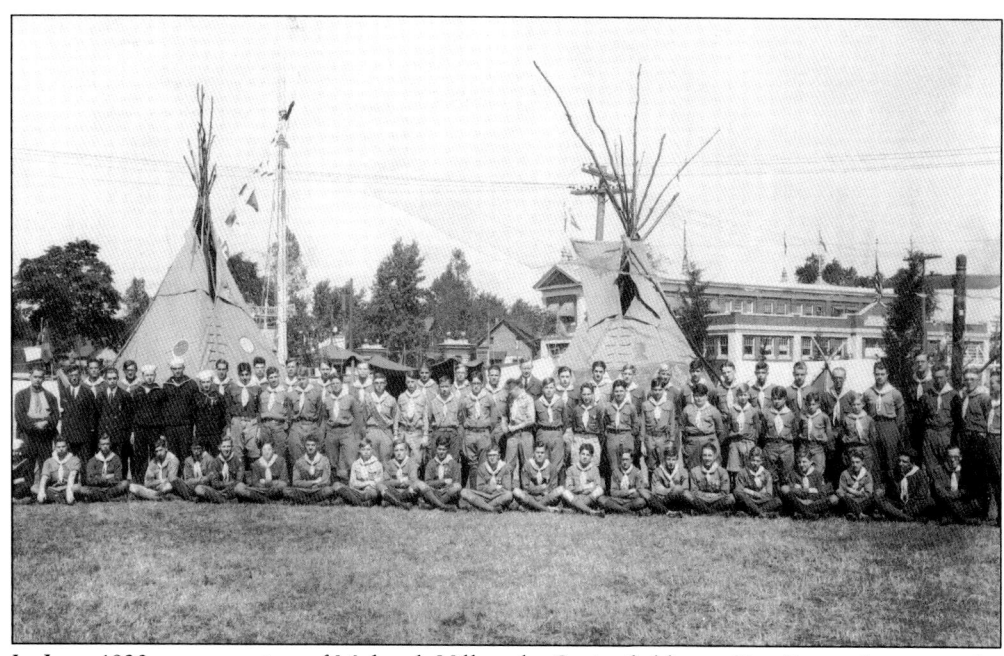

In June 1920, construction of Mohawk Village by Springfield-area Boy Scouts began in the northwestern corner of the exposition grounds. The long house and lodges were described in promotional literature as a "perfect reproduction of the Indian villages existing in the Eastern States prior to the advent of the White Man," with historical correctness "guaranteed by the active co-operation of Mr. A.C. Parker, New York State Archaeologist, himself an Iroquois Indian." Bronze medals were awarded to the Boy Scouts who worked the longest on the construction. The six lodges were occupied during the exposition by dozens of Scouts representing the 10 eastern states, from Maine to Pennsylvania. A garden "raised from genuine Mohawk seed and cultivated in the Indian's manner" was also part of the village.

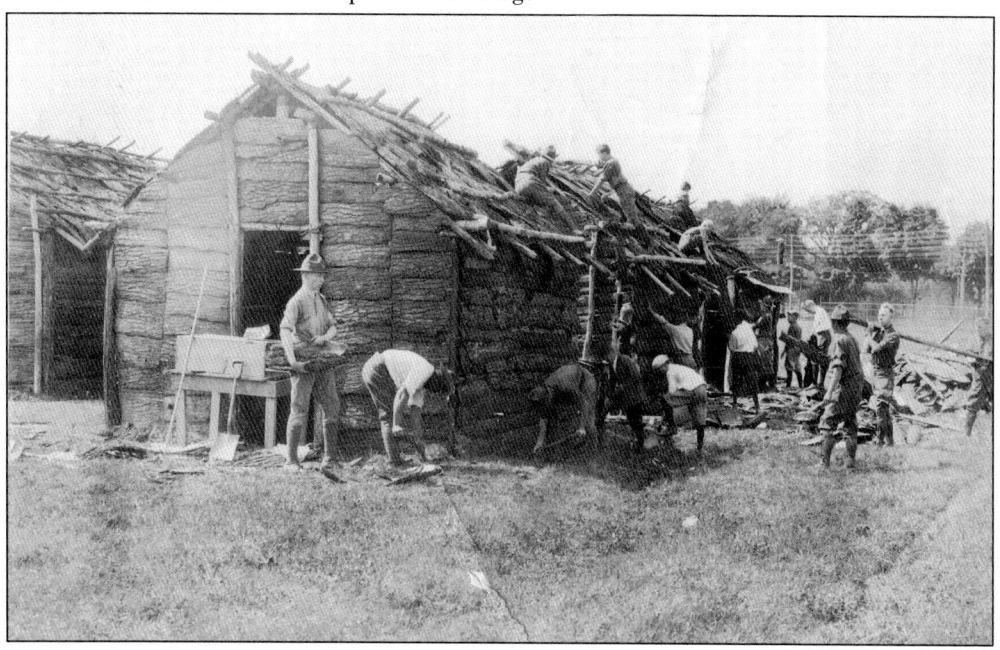

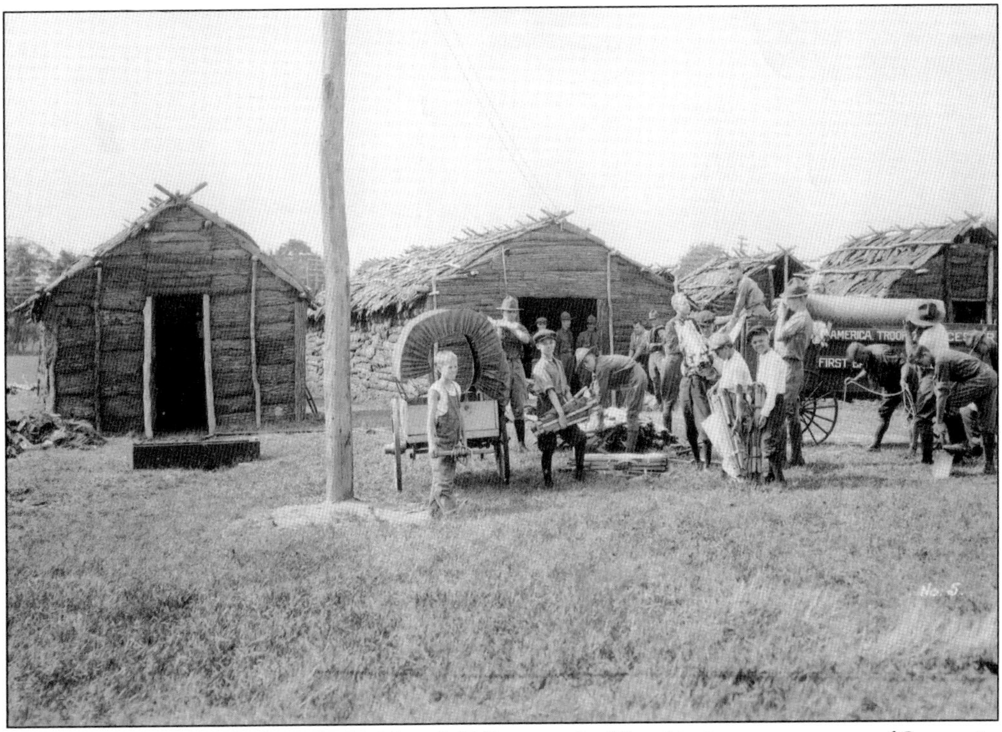

National scout officials described Mohawk Village as the "first big interstate event of Scouts in camp ever held in the United States." In its second year, a Scout camp was added to accommodate 80 additional Boy Scouts during the fair. Scouting leaders "pronounced the Village one of the most inspirational things accomplished for boys in this country," and within a short time, Scout-built Indian villages were appearing at fairs throughout the country. The bark siding and roofing of the structures deteriorated after several seasons, and the long house and lodges were replaced in 1926 with teepees in Mohawk Village, with full knowledge that teepees were used by Plains Indians and not Mohawks.

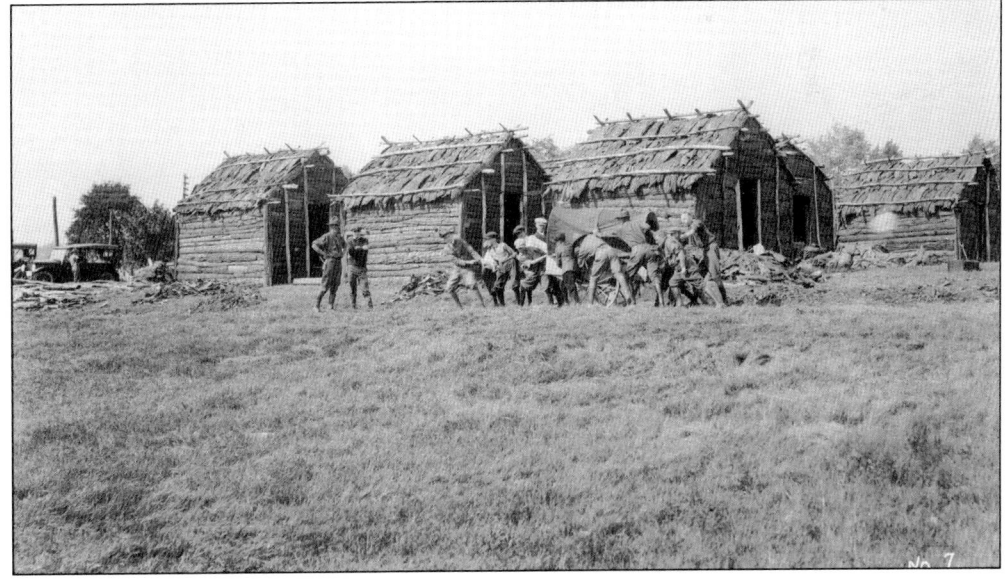

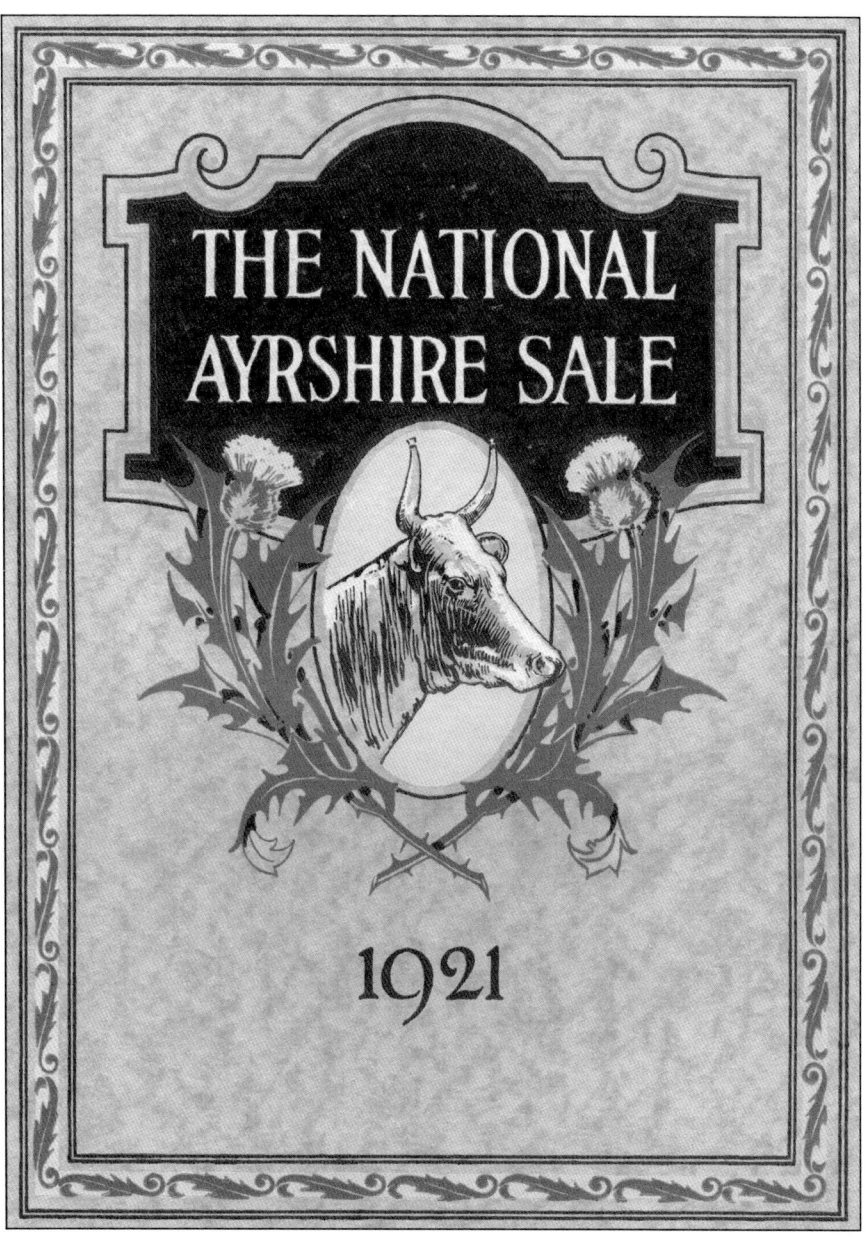

Another event held on the grounds of the exposition, for the third consecutive year, was the national sale of Ayrshire cattle on Tuesday, June 7, 1921. The impressive sale catalog features a stamped cover and more than 80 pages of livestock offerings, including Aldebaran Janet 54334, Auchincloigh Violet 12th 35267 (imported), Woronoake Nelli 42727, and Yellow Kate's Dandilette 60076. Consignors traveled from as near as Woronoco, Massachusetts (Horace A. Moses's Woronoake Heights), and Spencer, Massachusetts (Alta Crest Farm), and from as far as Wheeling, West Virginia (Hill Top Farm), Hudson, Ohio (Evamere Farm), and Waukesha, Wisconsin (Adam Seitz). The Sale Committee included a Mr. Stephen Bull—no joke! The annual meeting of the Ayrshire Breeders Association took place the next day at the Hotel Kimball in Springfield. Paul Reymann of Wheeling, West Virginia, was elected president, and the organization voted to quadruple its capitalization from $25,000 to $100,000. (Author's collection.)

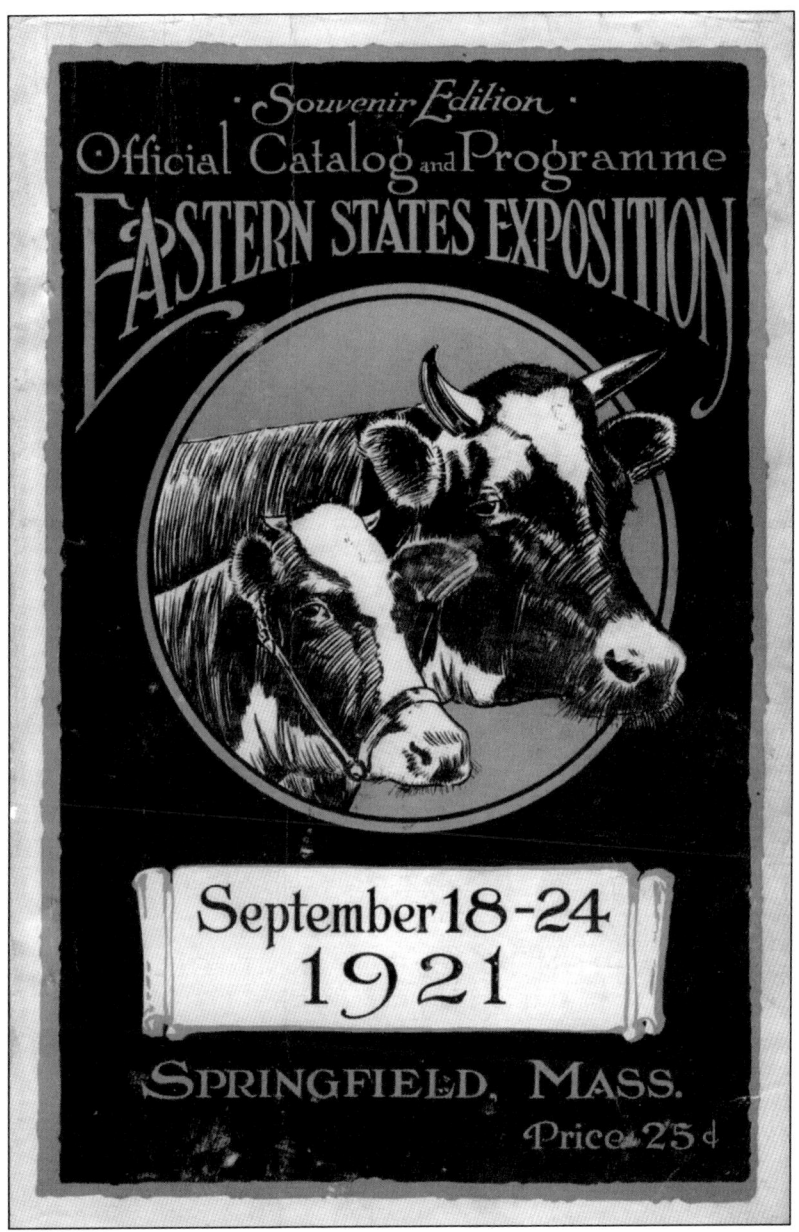

The cover of the exposition program was changed in 1921 to the layout pictured here; this design was used through 1924. It replaced the patriotic Uncle Sam theme that was used from 1917 to 1920. The program lists special days, including Music Day (Sunday, September 18), Farmers' and Governors Day (September 19), State of Maine Day (September 20), Dairy Cattle Day (September 21), Meat Producers' Day (September 22), Fruit and Garden Products and Children's Day (September 23), and Springfield and Automobile Day (September 24). The program also features livestock exhibitors, judging schedules, and listings of other activities. Among the advertisers are long-gone Springfield-area institutions such as Forbes & Wallace, Steiger's, Adams Nursery, Chapman Valve, and Cheney Bigelow Wire Works. The Charles C. Lewis Company of Springfield is one of two advertisers listed still in business today under the same name; the other is Hartford Insurance. (Author's collection.)

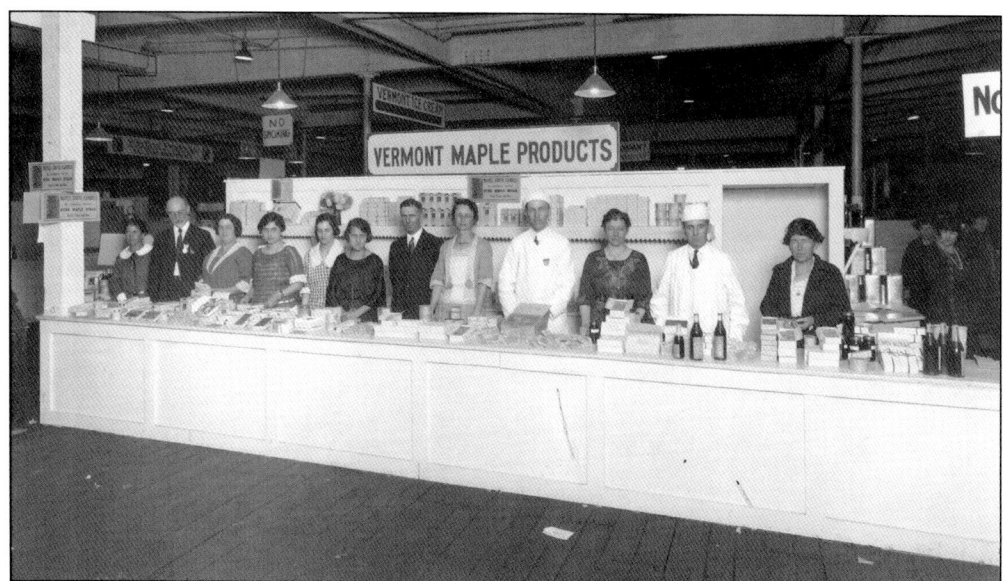

Then as now, Vermont maple products were a favorite of fairgoers. Before the construction of the Vermont State Building, Vermont exhibits were located outside the Massachusetts Building and inside Machinery Hall (now C Barn), as seen here. Staffed by a group that seemingly takes their work very seriously, the sweet offerings—maple syrup and maple sugar candy—remain the same.

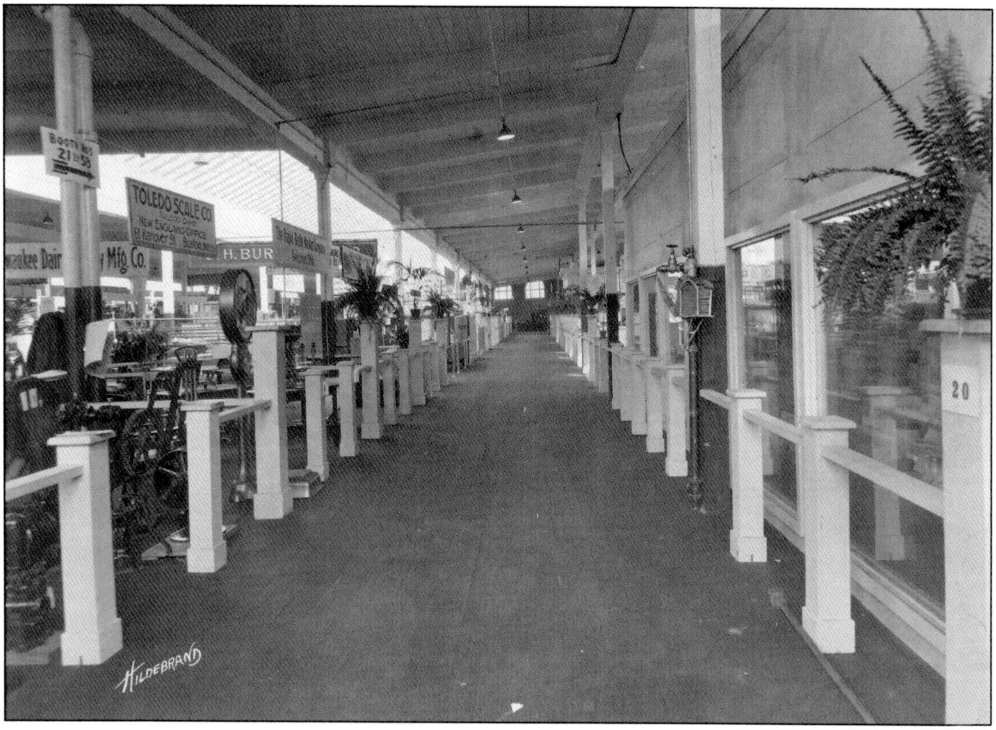

The Eastern States Exposition was promoted to exhibitors as a showcase "Where Manufacturer and Dealer Meet a Buying Public." Labor-saving devices for farm and home were on display throughout the fair. This image shows Machinery Hall (now C Barn), the location for these exhibits before the construction of the Industrial Arts Building.

This view is looking east from the Massachusetts Building. In 1922, the Girl Scouts' headquarters moved from a small encampment behind the Mohawk Village to the spacious log house seen at upper right, a location described in exposition literature as "one of the most auspicious and prominent locations on the grounds." Swimming demonstrations were given by Scouts in a wood and canvas swimming pool, not visible in this photograph. Again, note the litter!

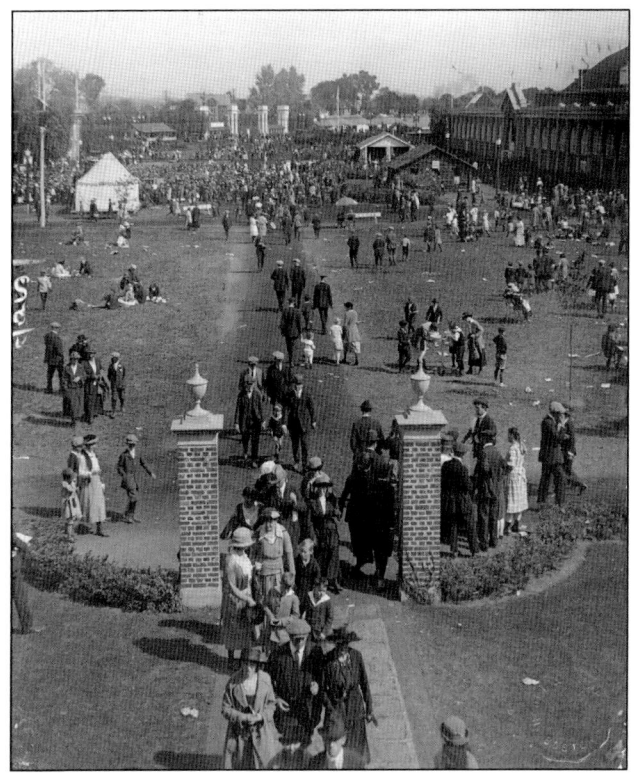

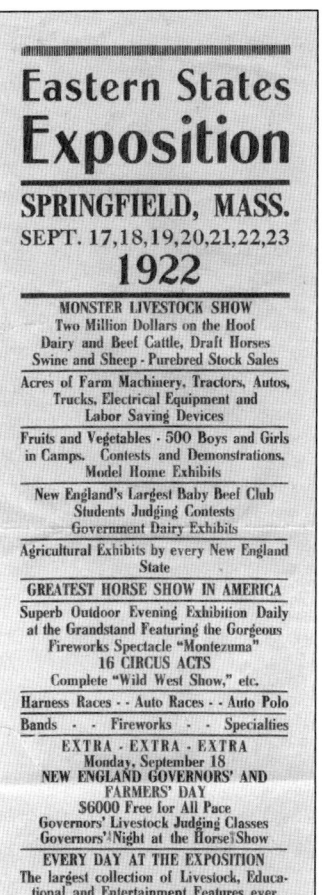

If the promise of the exposition's "largest," "greatest," and "superb" attractions were not enticing enough, railroads offered reduced fares from all stations to attract passengers. The fireworks spectacle Montezuma, referenced in this broadside, was promoted as "the largest and most pretentious offering of its kind ever attempted in America. Tons of fireworks and high explosives are used to produce the wonderful effects that continue for more than two hours." (Author's collection.)

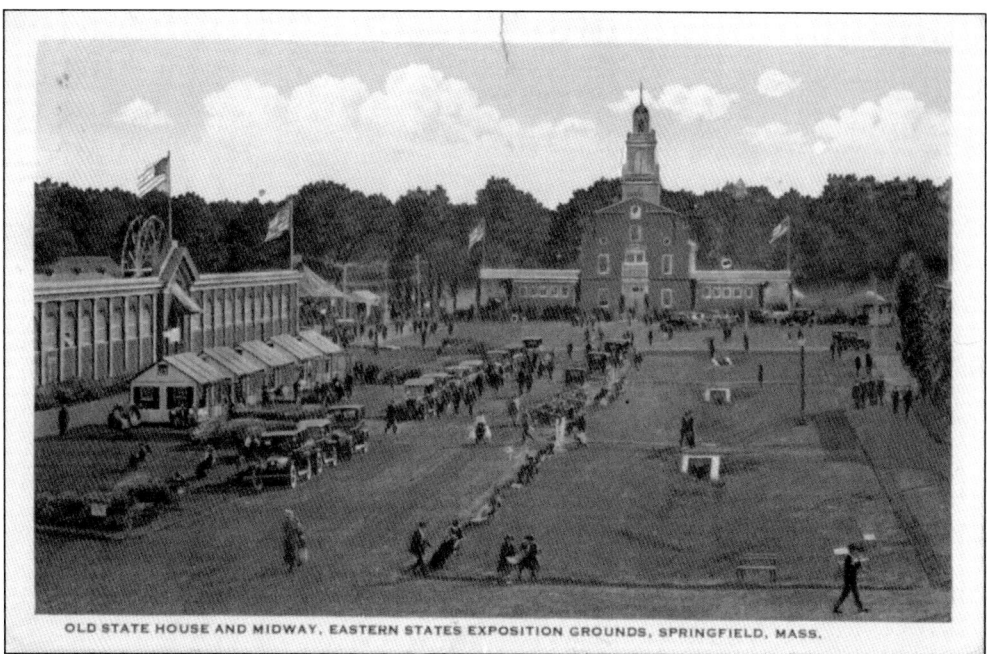

OLD STATE HOUSE AND MIDWAY, EASTERN STATES EXPOSITION GROUNDS, SPRINGFIELD, MASS.

The exposition's 1920 rules and regulations state, "Fast driving or riding within the Exposition grounds is strictly prohibited. Any violation of this rule will incur severe penalty. Obstruction of any road, path or passage with automobile or other vehicle is forbidden under penalty of expulsion from the grounds." Automobiles are no longer allowed on the grounds during open hours. (Author's collection.)

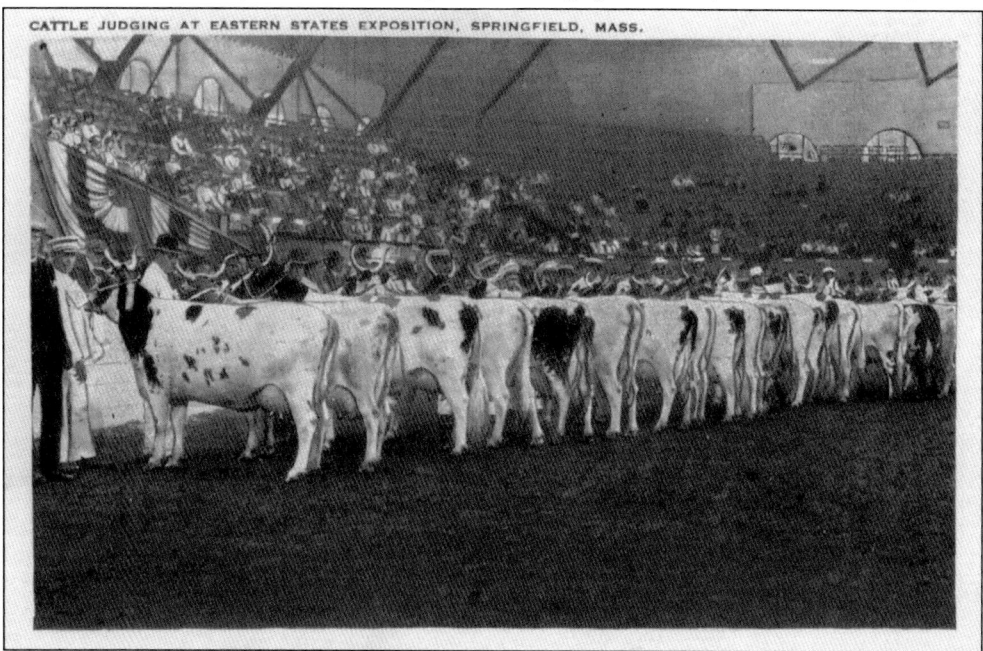

CATTLE JUDGING AT EASTERN STATES EXPOSITION, SPRINGFIELD, MASS.

In 1922, the amount offered in premiums in the cattle category was more than $28,000. In order to encourage better dairy farming and breeding in Massachusetts, the Department of Agriculture additionally awarded gold medals to the best Holstein, Jersey, Ayrshire, Guernsey, and Milking Shorthorn cows bred and owned in Massachusetts and shown at the exposition. (Author's collection.)

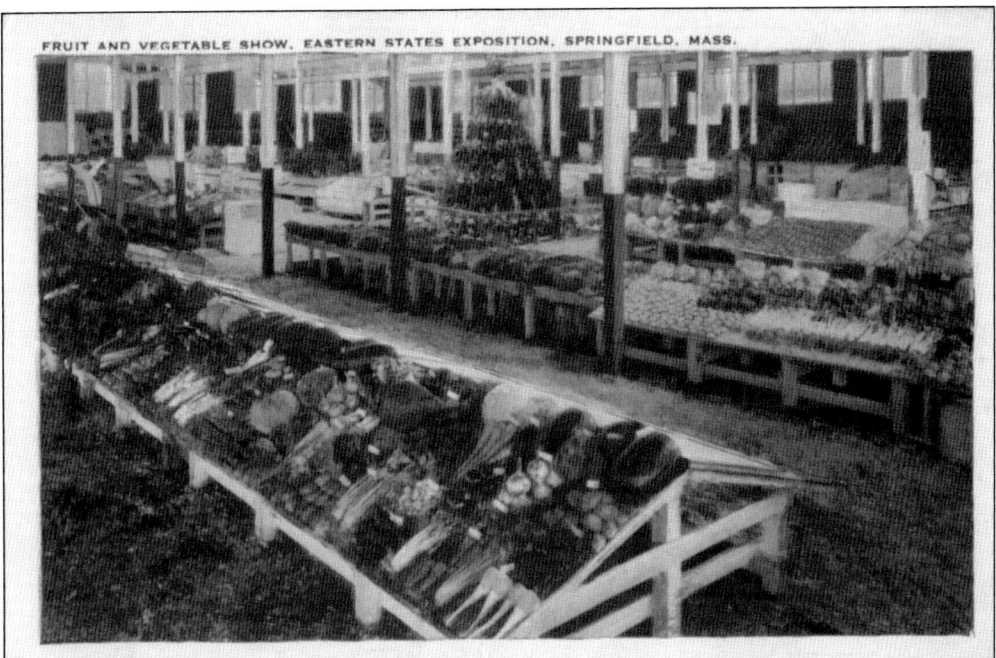

The purpose of the fruit and vegetable display at the exposition was to educate the general public about the high quality of produce grown in this region and encourage the greater consumption of locally grown fruits and vegetables. (Author's collection.)

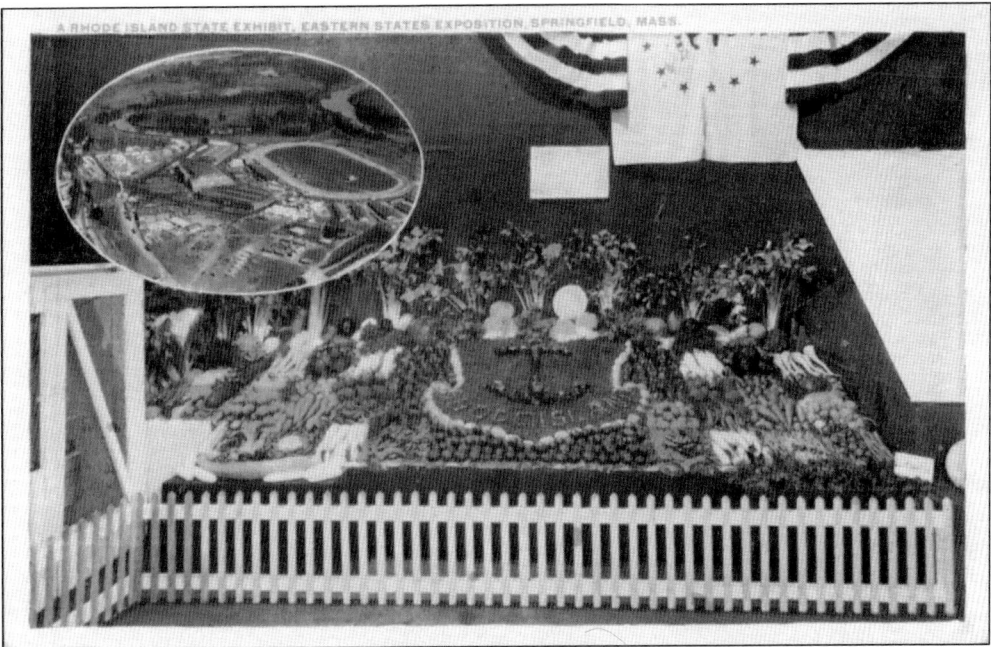

A special vegetable display accompanied a Rhode Island state exhibit highlighting three of the Ocean State's most famous agricultural products—the Rhode Island Greening apple, the Rhode Island Red hen, and Rhode Island white-cap corn, of the variety used to make Rhode Island Johnny cake meal. (Author's collection.)

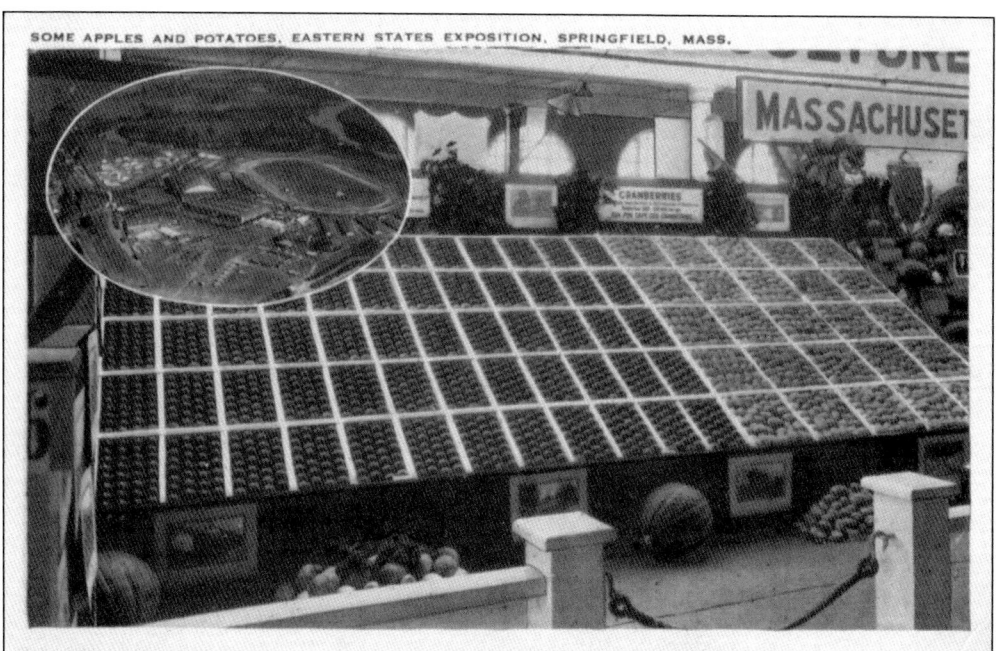

More than 35,000 apples were distributed on Wednesday, September 19, 1923—Eastern States Apple Day. The promotion was designed to remind fairgoers of the high quality of New England fruit and encourage growers to continue the restoration of the region's orchards, which had been allowed to deteriorate since the Civil War. (Author's collection.)

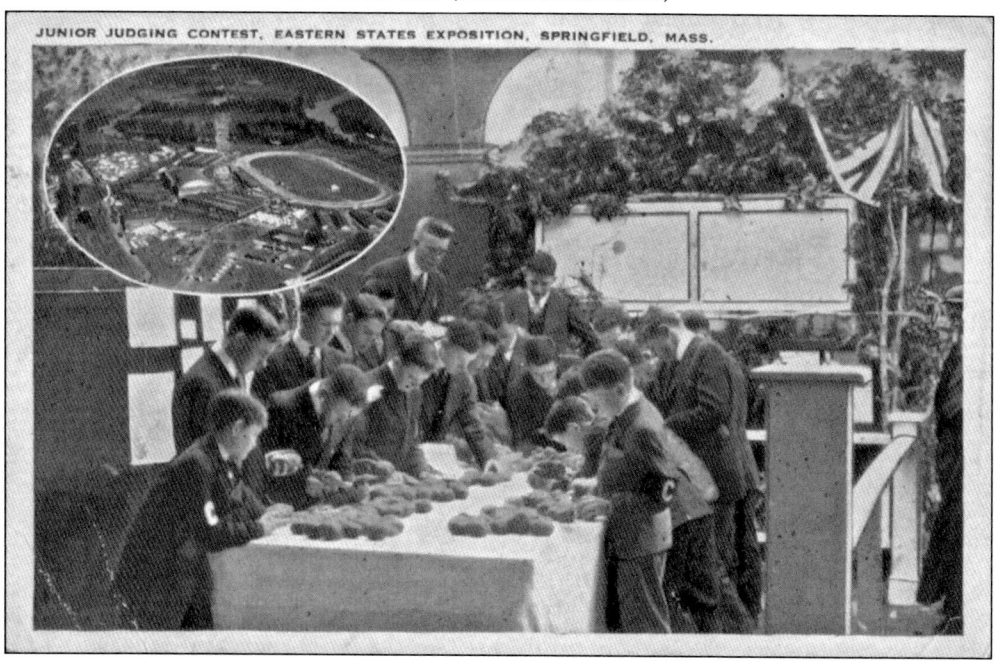

Youth activities were always an important part of the exposition. The Boys' and Girls' Department consisted of Camp Vail, Junior Achievement Village, Mohawk Village, Girl Scout Camp, and the Baby Beef Club Camp. The exposition billed its youth offerings as "educational and entertaining— the greatest junior achievement in the United States." (Author's collection.)

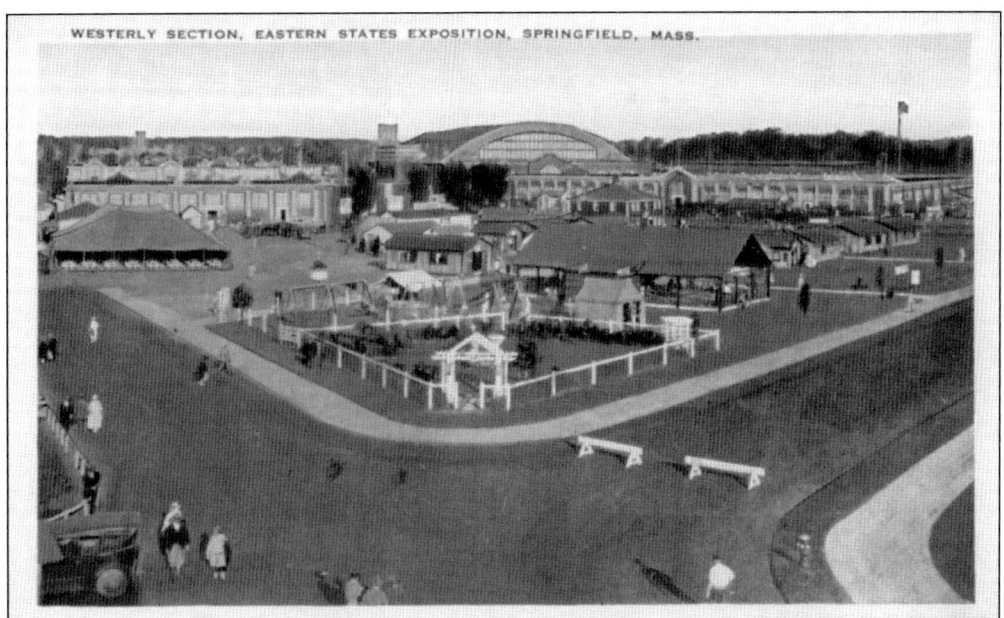

According to an exposition Home Department pamphlet, "the sphere of women in the home has not been neglected." The Home Department consisted of eight cottages, a large "assembly tent," and a model playground located north of the Massachusetts Building on land where additional state structures would later be constructed. The Home Department hoped to provide every woman a "broader, better outlook on life." (Author's collection.)

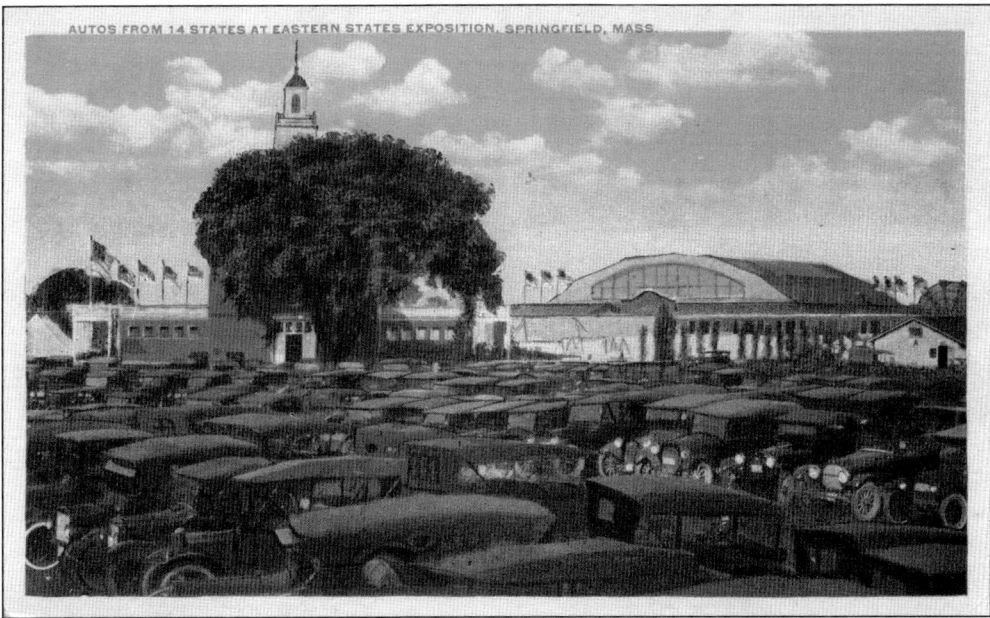

Many visitors arrived via automobile, and before the opening of the Hampden County Memorial Bridge in 1922, more than a few of those vehicles utilized the old wooden toll bridge that it replaced. On Friday, September 23, 1921, a total of 1,123 trucks and buses, 9,484 touring cars, 1,935 runabouts, and 180 motorcycles crossed the bridge, along with 8,586 pedestrians and 411 horse-drawn vehicles. (Author's collection.)

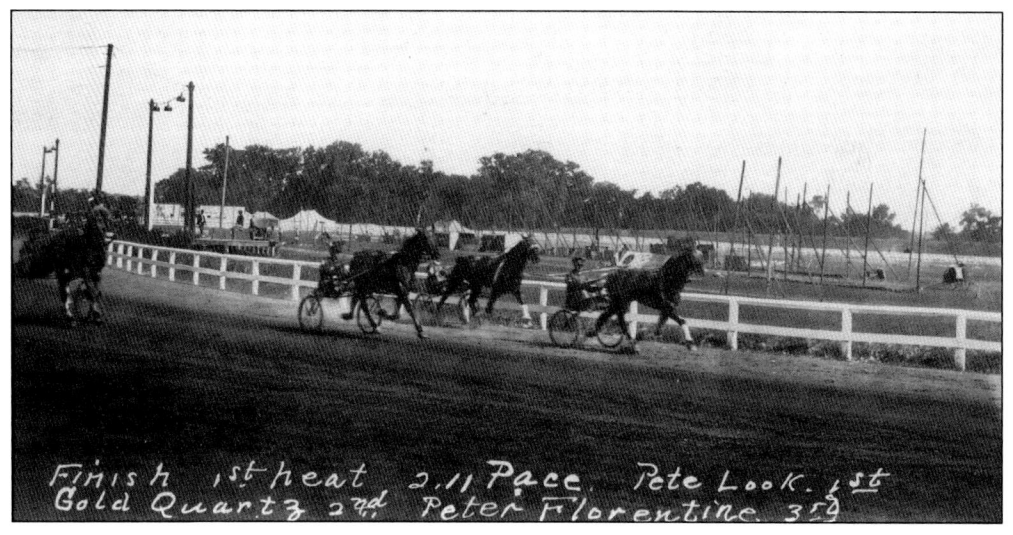

A note dated September 26, 1923, from Vernet H. Keller of the Hampshire Manufacturing Company in Hatfield, Massachusetts, to Fred L. Yale of Meriden, Connecticut, accompanied these photographs and read: "As per promise made you in our very pleasant visit at the gate to the race just before the great race at The Eastern States afternoon of the 17th I am enclosing some snaps of the Free-for-all and the 2.11 pace. Haven't seen any account in the papers of any one from your city being kidnapped so presume you reached home safely. With kindest regards I am . . . Yours Truly, V.H. Keller." (Both, author's collection.)

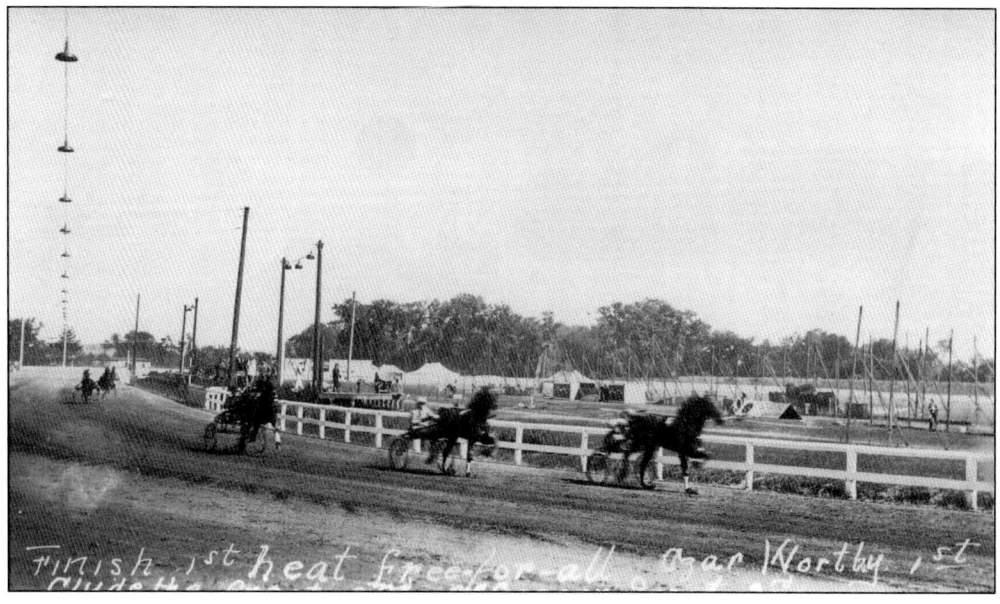

The local influence of the exposition cannot be underestimated. The cover for this booklet about Exposition View in neighboring Agawam includes the view from across the river where the development was located. Described as the "million-dollar development," its location "overlooking" and "three minutes from" the grounds of the exposition was prominently touted. More than 500 people attended an open house at the Reed Street location. (Author's collection.)

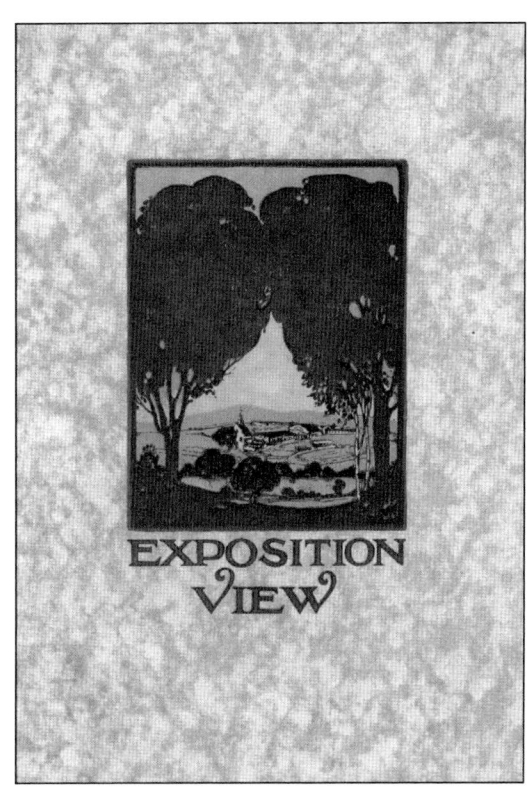

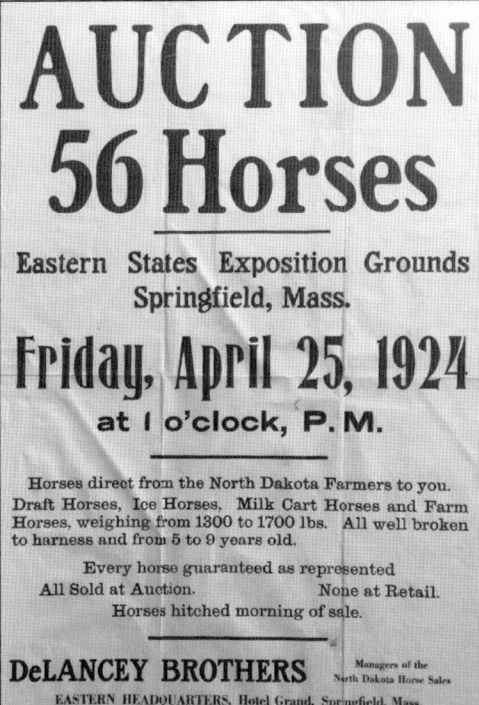

This poster advertises one of several horse auctions conducted on the exposition grounds by DeLancey Brothers of North Dakota Horse Sales. Other auctions took place in February, March, and May. Though demand would soon drop dramatically, in 1924, there was still enough of a market for "farm chunks," milk cart, ice, and draft horses to make these auctions worthwhile. (Author's collection.)

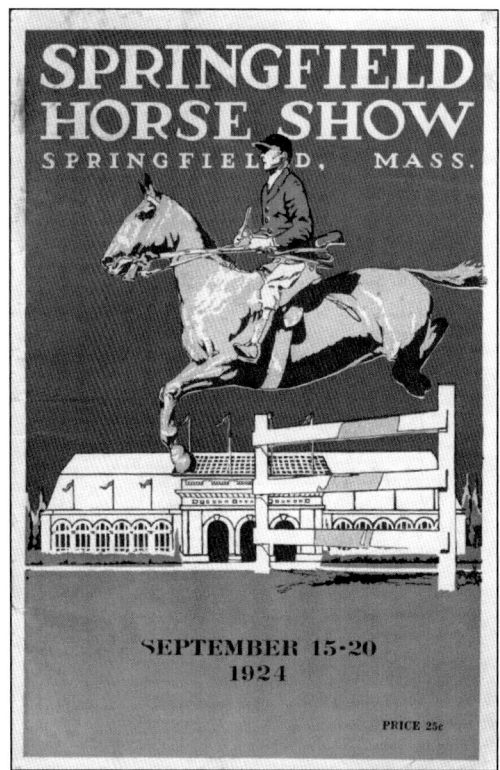

The Springfield Horse Show has been held at the exposition since 1916. It was renamed the Eastern States Horse Show in 1940. A more formal affair in years past, show programs included advertisements from the London Harness Company, B. Altman, Brooks Brothers, and, before the stock market crash, many investment bankers. (Author's collection.)

Occupying the back cover of the Springfield Horse Show program for several years in the 1920s were advertisements from Rolls-Royce, which were manufactured in Springfield from 1920 until 1931, the only place outside of England that the luxury vehicles have ever been built. Once the Great Depression set in, advertisements for Lux Soap and Dutchland Farms ice cream took Rolls-Royce's place on the back cover. (Author's collection.)

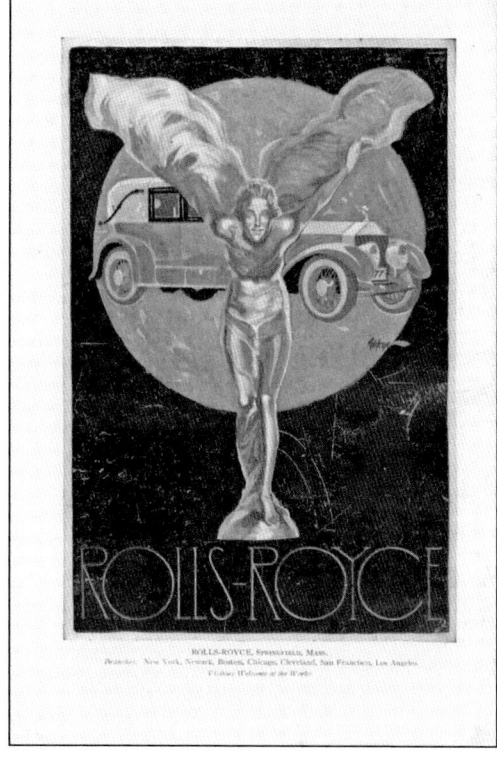

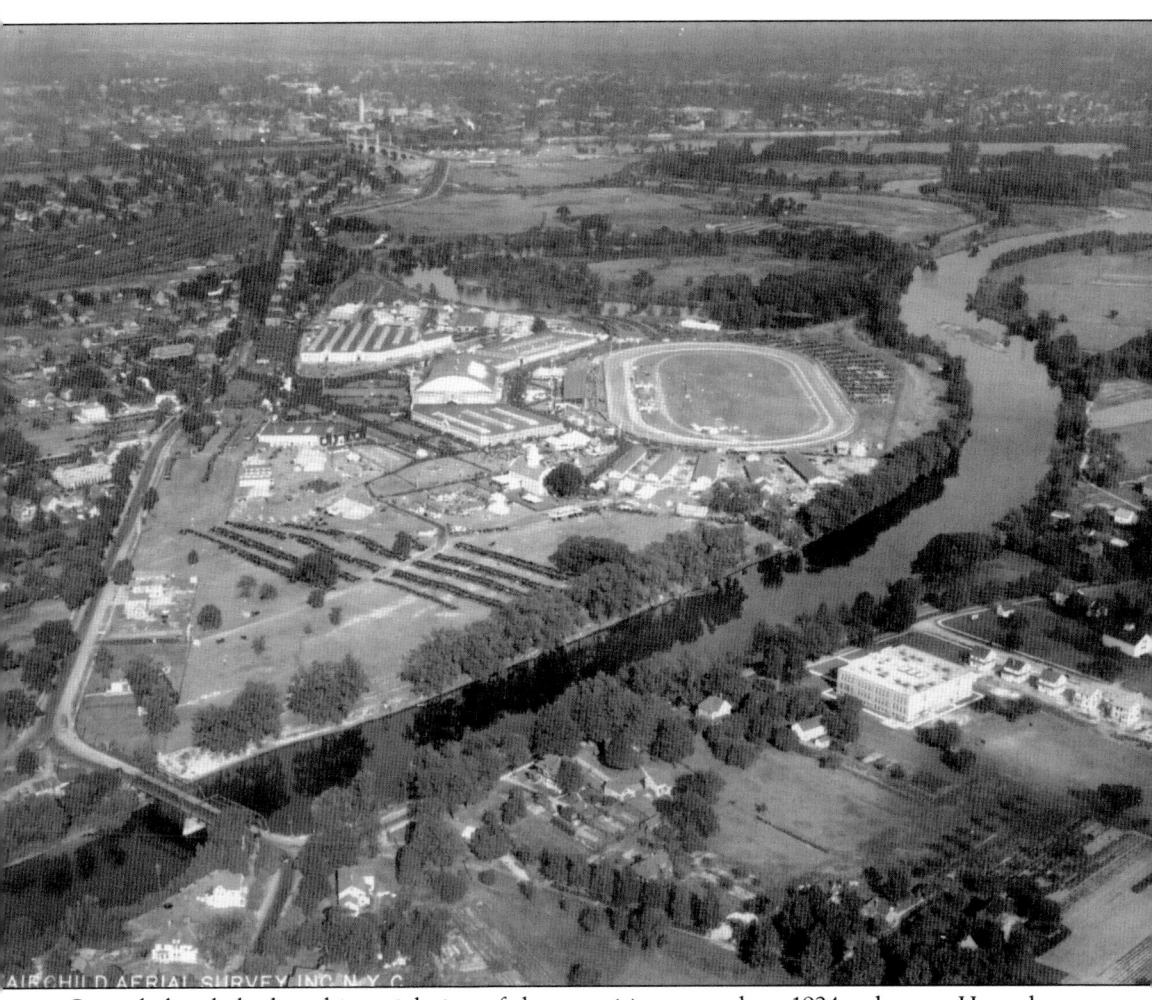

Several clues help date this aerial view of the exposition grounds to 1924—the new Hampden County Memorial Bridge, top center, and the new Agawam High School (now the Roberta G. Doering School), bottom right, were both dedicated in 1922; the five houses under construction just beyond the school are part of the Exposition View development that was begun in 1924; that same year, the exposition's new Industrial Arts Building opened, seen just beyond and to the left of the Coliseum, to the left of the racetrack at center. These were years of expansion for the exposition, and the next year would see even more changes to the fairgrounds with the construction of the Junior Achievement Building, the Hampden County Improvement League Building, and the Maine Building. The open fields directly beyond the fairgrounds are now home to several shopping centers and industrial parks.

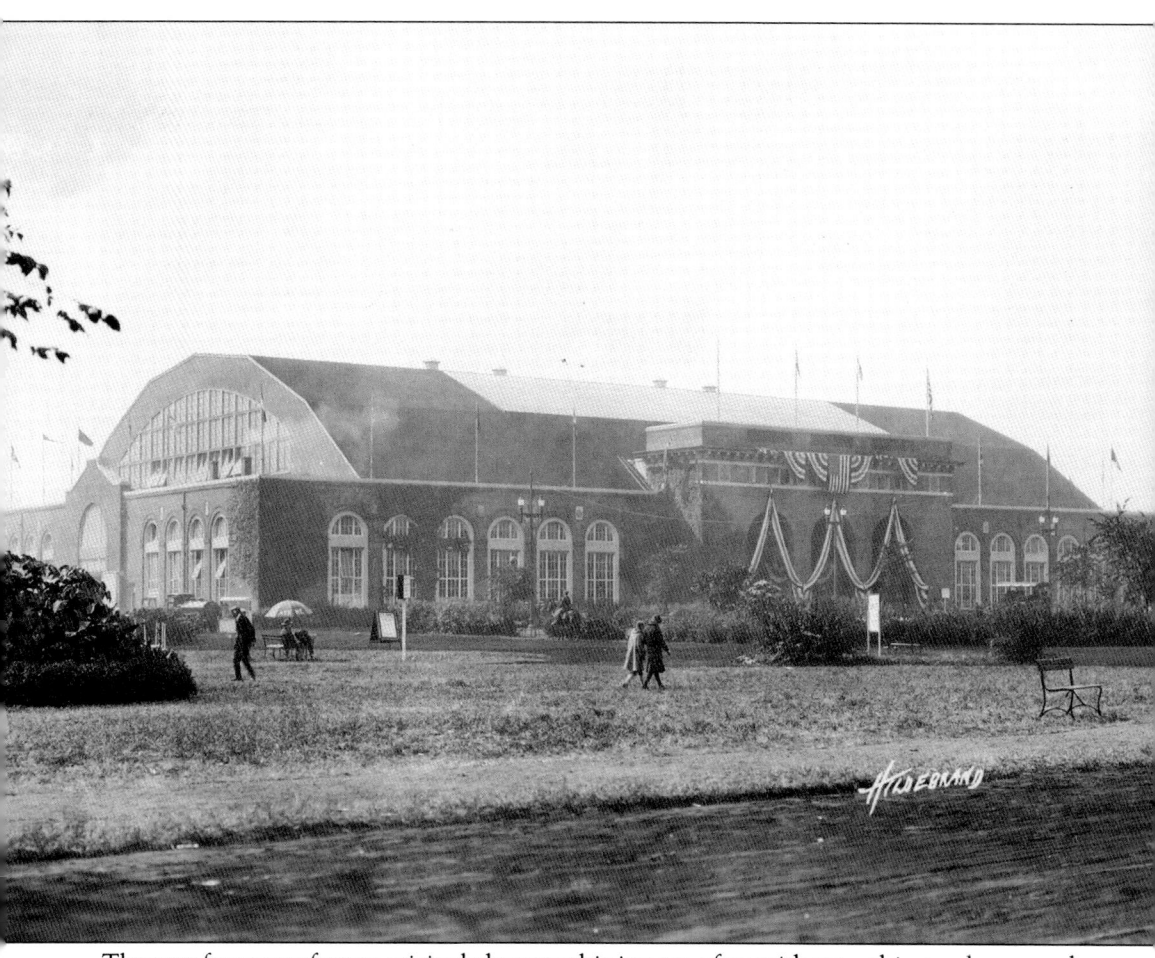

The next four pages feature original photographic images, often with retouching and crop marks visible, that were used to produce a series of postcards in the 1920s. Some subtle (and sometimes not so subtle) changes were incorporated to create the finished product. The Coliseum is seen here, patriotically decked out in bunting and flags. The flags all snap in a stiff breeze on the finished postcard. The umbrella at left center imprinted with "Ask Me" marked one of many information stations located throughout the grounds and staffed by Boy Scouts, who would answer more than 26,000 questions from visitors during the course of the fair. Also seen at center is a rider on horseback, who did not appear on the postcard.

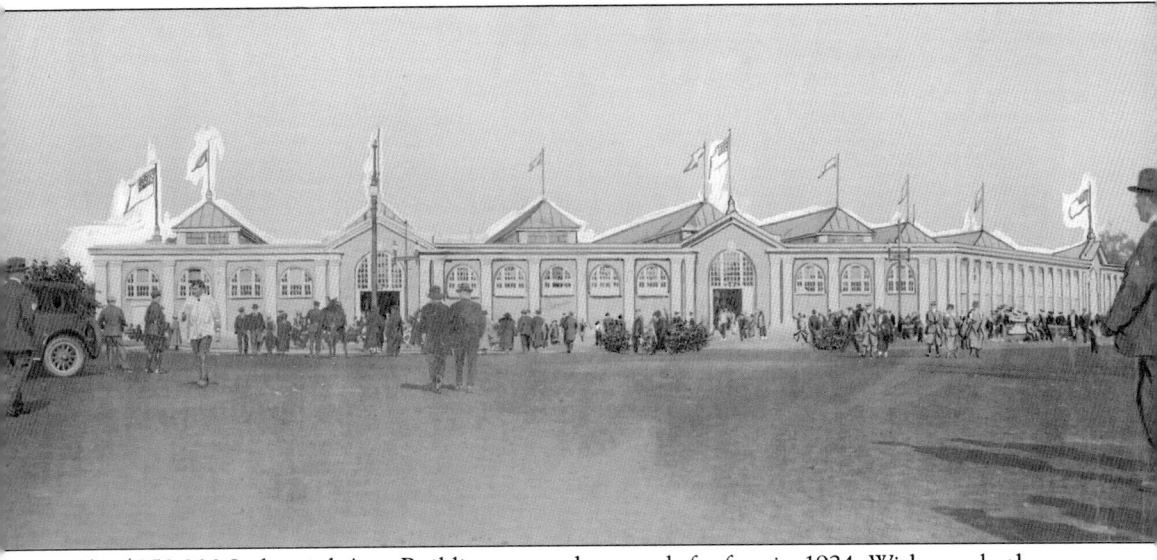

The $250,000 Industrial Arts Building opened to much fanfare in 1924. With nearly three acres of floor space, it was for decades the largest structure of its type at any fair or exposition in the United States. Exposition promotional material boasted that "none but the highest grade industrial concerns and educational institutions permitted to exhibit—this ensures—'America's Finest Industrial Exhibition.' " A partial list of exhibitors from its opening year included several Springfield area concerns: American Bosch, American Saw, Baker Extract, Kibbe Brothers Candies, Rolls-Royce of America, Albert Steiger, Orr Motor Company, and Tait Brothers Dairy. Nearly a century later, the Industrial Arts Building, now known as the Better Living Center, still plays host to a variety of trade shows every year.

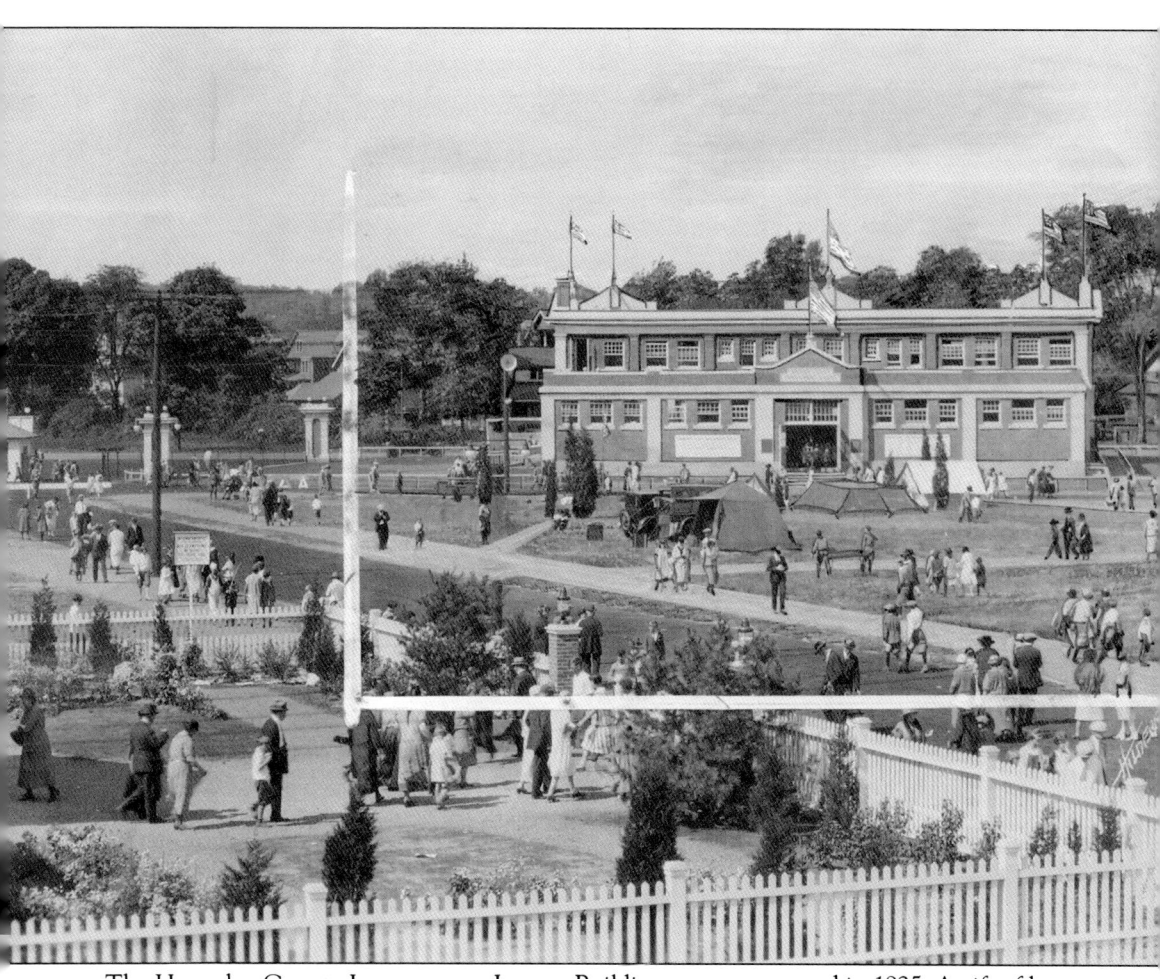

The Hampden County Improvement League Building was constructed in 1925. A gift of league founder Horace A. Moses, the structure cost nearly $75,000 and would serve as the home of the league for many years. Still active more than a century after its founding, the organization continues its efforts to make Hampden County a better place to live and work by supporting agriculture, education, and 4-H. The league's scholarship program supports higher education for young people from an agricultural background and others pursuing a career in agriculture. The league also provides grants for agricultural projects within the county and supplies equipment and assistance for embryology programs in local schools.

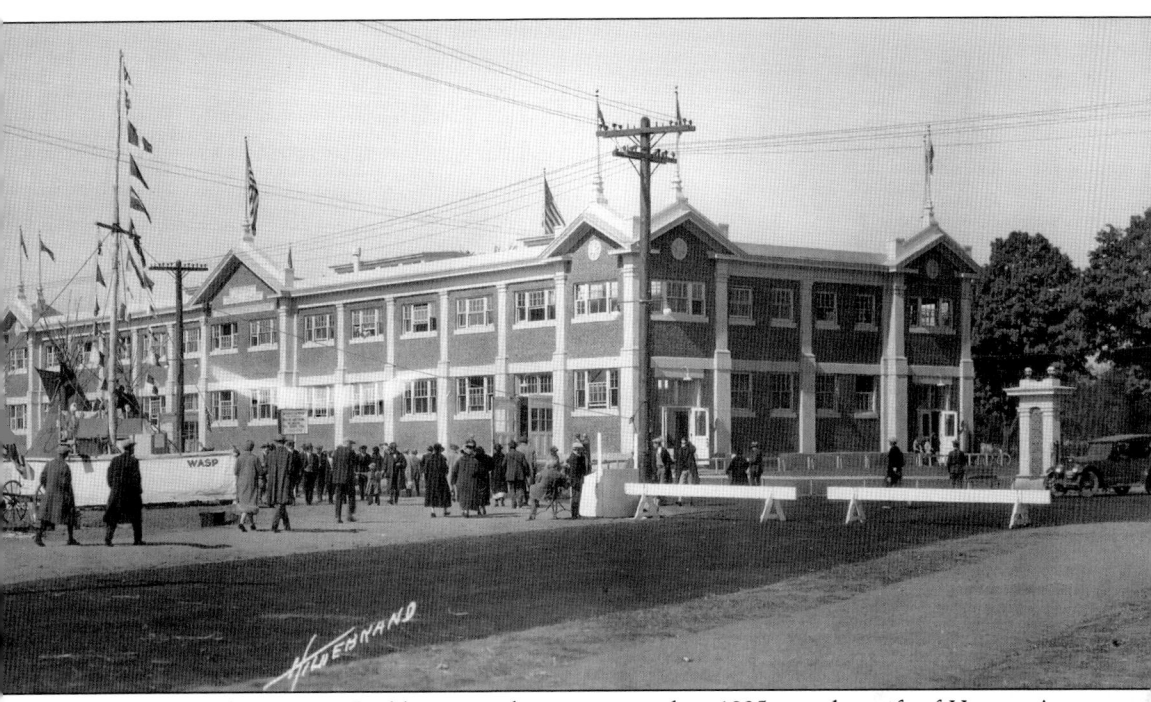

The Junior Achievement Building was also constructed in 1925, another gift of Horace A. Moses, one of the founding members of the Junior Achievement movement. The large two-story building cost $100,000 and features a 500-seat auditorium, an exhibition hall, and a dormitory accommodating 600. Founded in 1919, the Junior Achievement organization was originally regional in nature and headquartered in Springfield, Massachusetts. The organization grew rapidly and by 1942 had spread nationwide. When the headquarters were moved to New York City that year, the West Springfield building was turned over to the exposition. In 1949, it was rededicated as the Horace A. Moses Memorial Building. The "landship" *Wasp*, seen at left, was a project of the Sea Scouts.

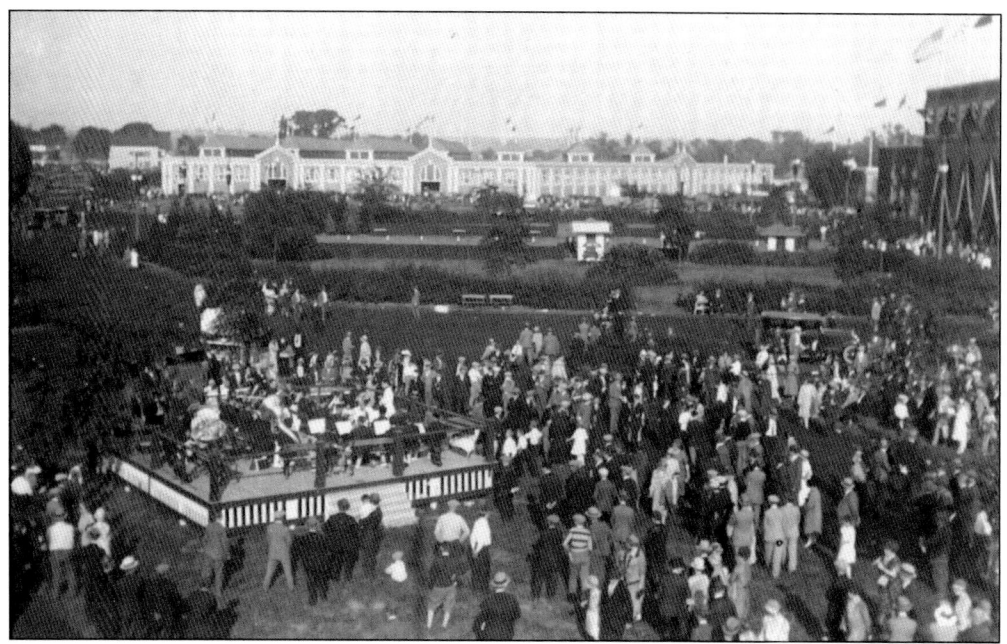

Music Day was a staple of the exposition for several years. The daily musical program consisted of concerts in the Coliseum, the racetrack grandstand, the Massachusetts Building, and the bandstand in the front of the Coliseum, where Victor's Concert Band often performed.

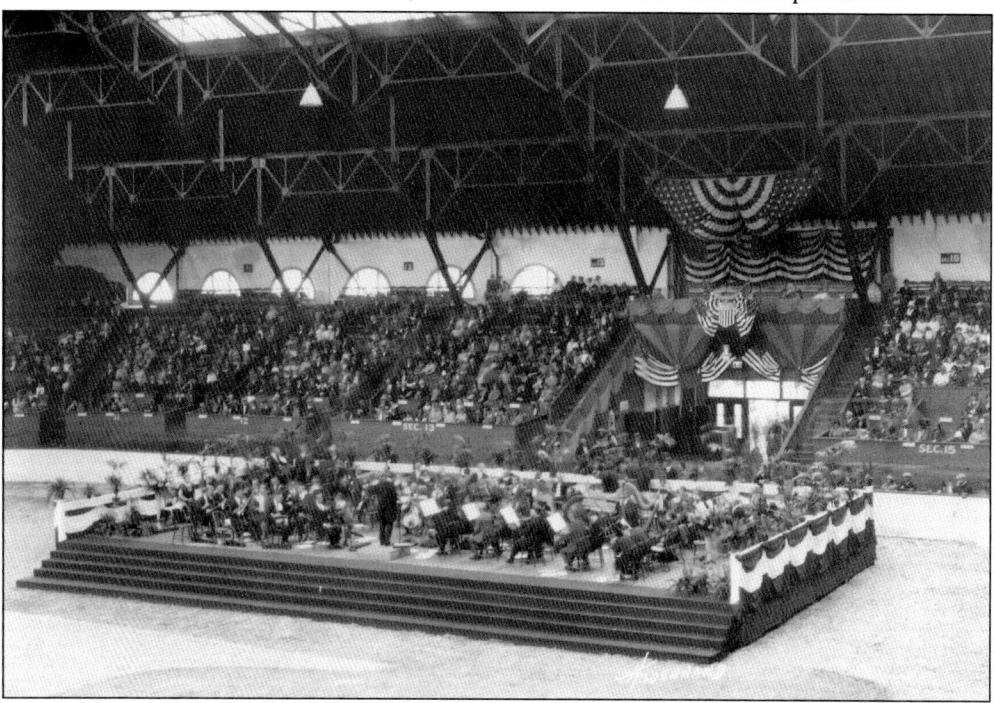

The Springfield Municipal Orchestra was the featured ensemble at the evening Music Day concert that took place in the Coliseum. John Philip Sousa and the New York City Police Department Band were among other musical groups invited to play over the years. Music continues to occupy an important role at the exposition.

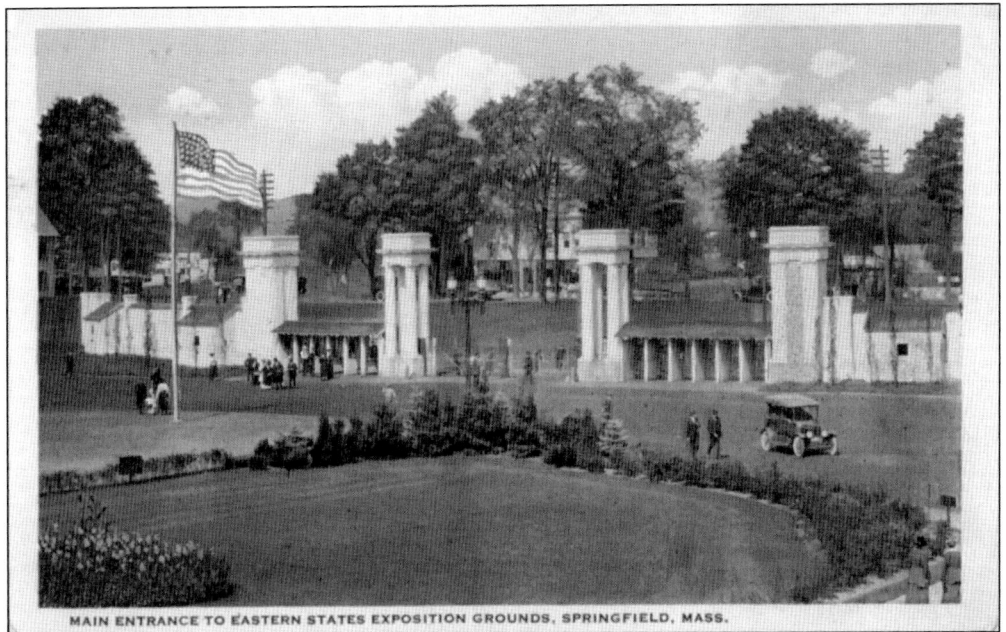

The main entrance to the exposition has never moved but was originally located on New Bridge Street. Renamed Memorial Avenue in 1926, the George Washington Memorial Highway was formally dedicated on June 8, 1932, in commemoration of Washington's 1775 route through the town on his way to Boston to assume command of the Continental Army. The ceremony was followed by a luncheon in the Coliseum. (Author's collection.)

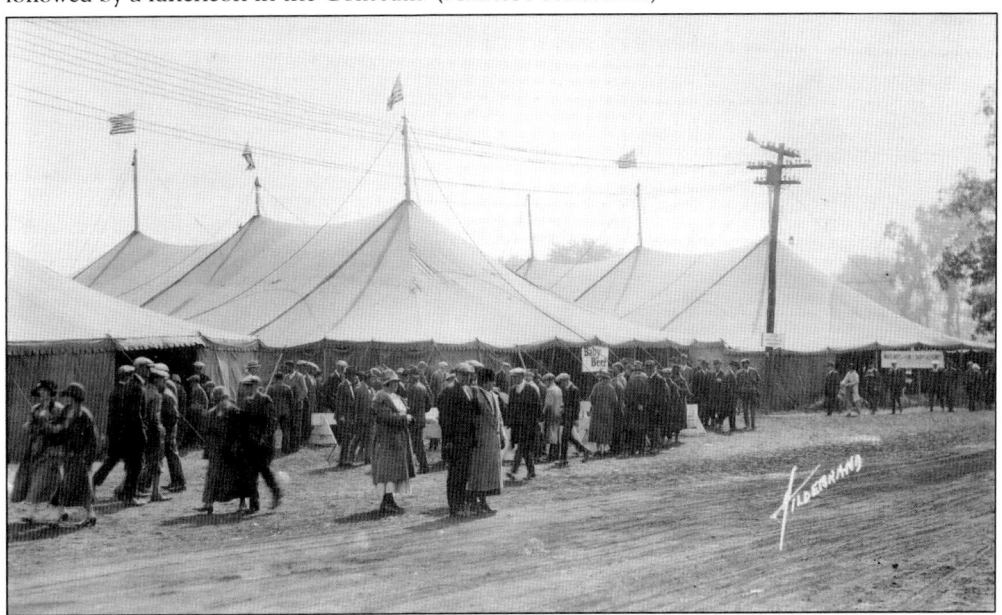

The Baby Beef Club Camp, housed in this tent, was introduced in 1920 and described by the exposition as "unique in American exposition history." To encourage the growing of native stock, 70 baby beeves were distributed to be raised by Hampden and Hampshire County youngsters. Thrift and industry were also learned in the feeding and management of the beeves. A total of $900 in premium money was offered the first year of judging.

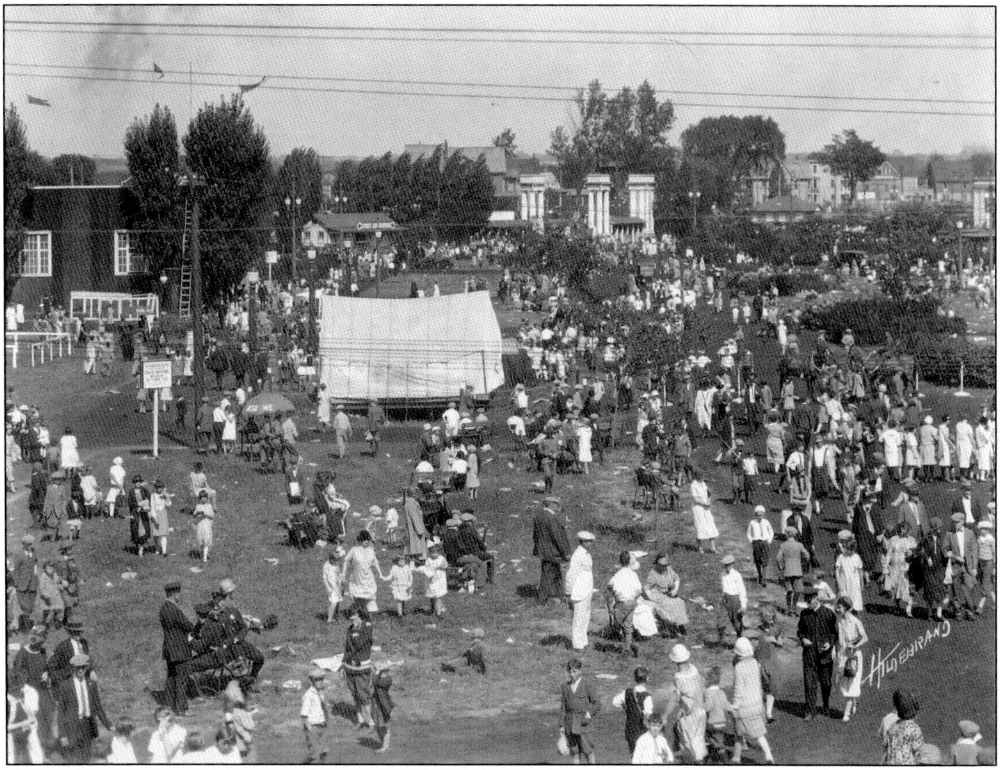

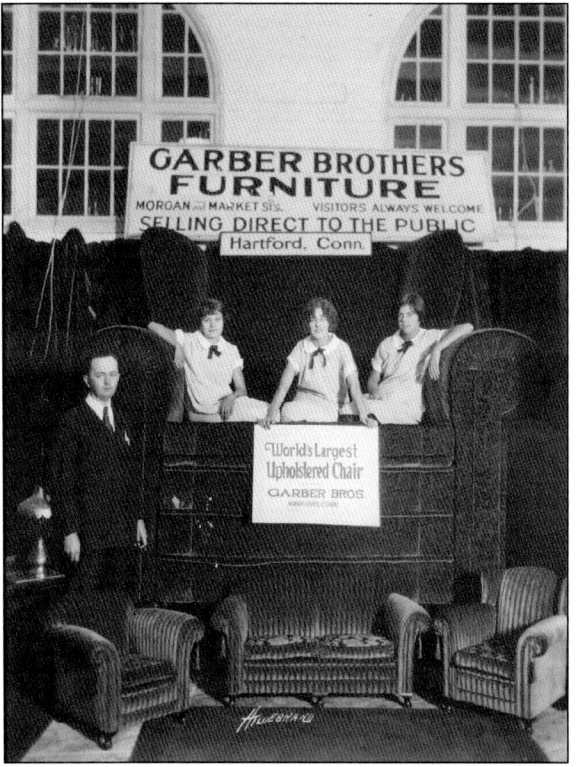

Looking east from the Massachusetts Building, note the "Ask Me" information station at center left. To anyone who grew up listening to Woodsy Owl tell a generation to "Give a Hoot–Don't Pollute," the amount of litter on the ground seems remarkable. Thankfully times have changed, and despite the tons of trash generated daily, the overall impression of today's fair is *not* of litter-strewn grounds.

More than a quarter million visitors attended the 1925 exposition, many of them crowding the Industrial Arts Building, no doubt to see firsthand the Garber Brothers exhibit featuring the largest upholstered chair in the world, manufactured by the Hartford, Connecticut, furniture makers.

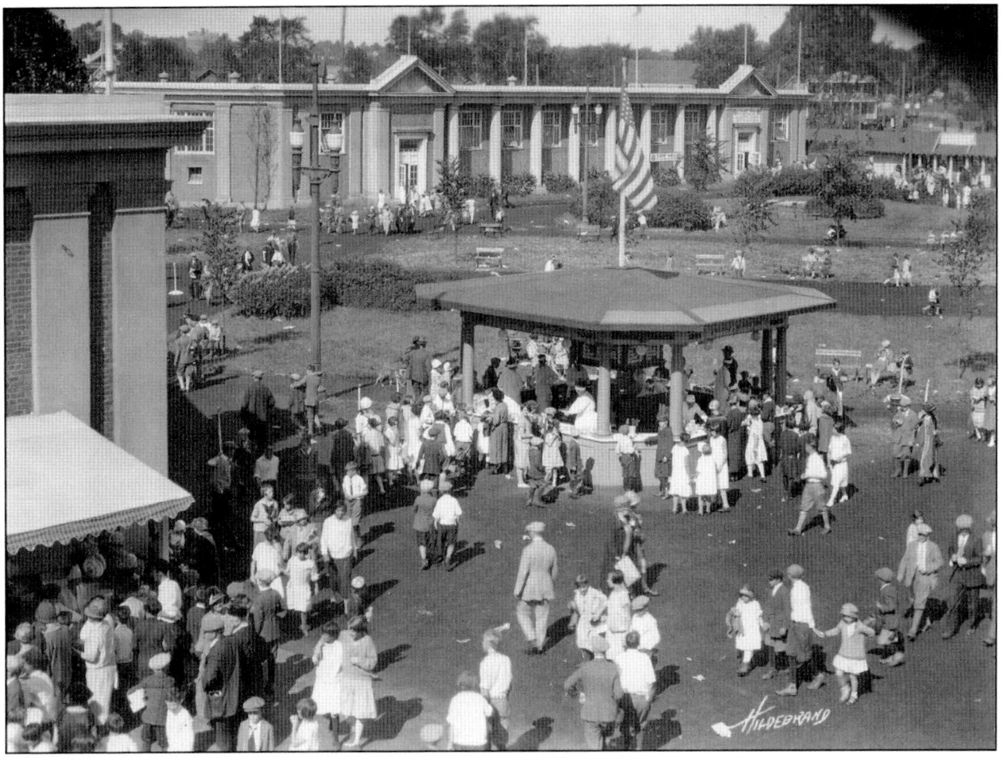

This view is looking northwest toward the Women's Building—at the time, home of Camp Vail—and currently the New England Center. At left is the former Machinery Hall, which was remodeled into a horse and poultry barn (now C Barn) upon the opening of the Industrial Arts Building. The refreshment stand at center was operated by the Waldorf System, an early restaurant chain, and offered ham or cheese sandwiches for a dime; roast beef sandwiches cost 15¢.

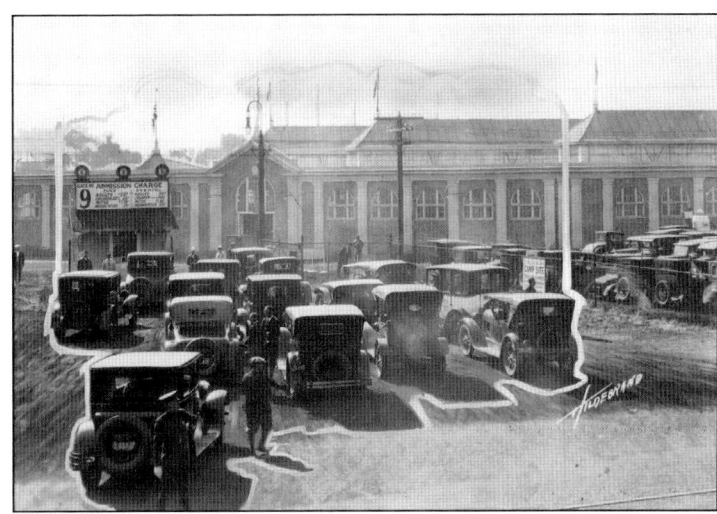

The brochure cover on which this retouched photograph appears notes, "Autos entering grounds from 34 states." Much was made of the broad participation in and broad appeal of the exposition. A pamphlet for the Industrial Arts Building, seen in the background, touts its "roof of monitor construction" clearly visible. This feature was lauded for its "excellent light and ventilation."

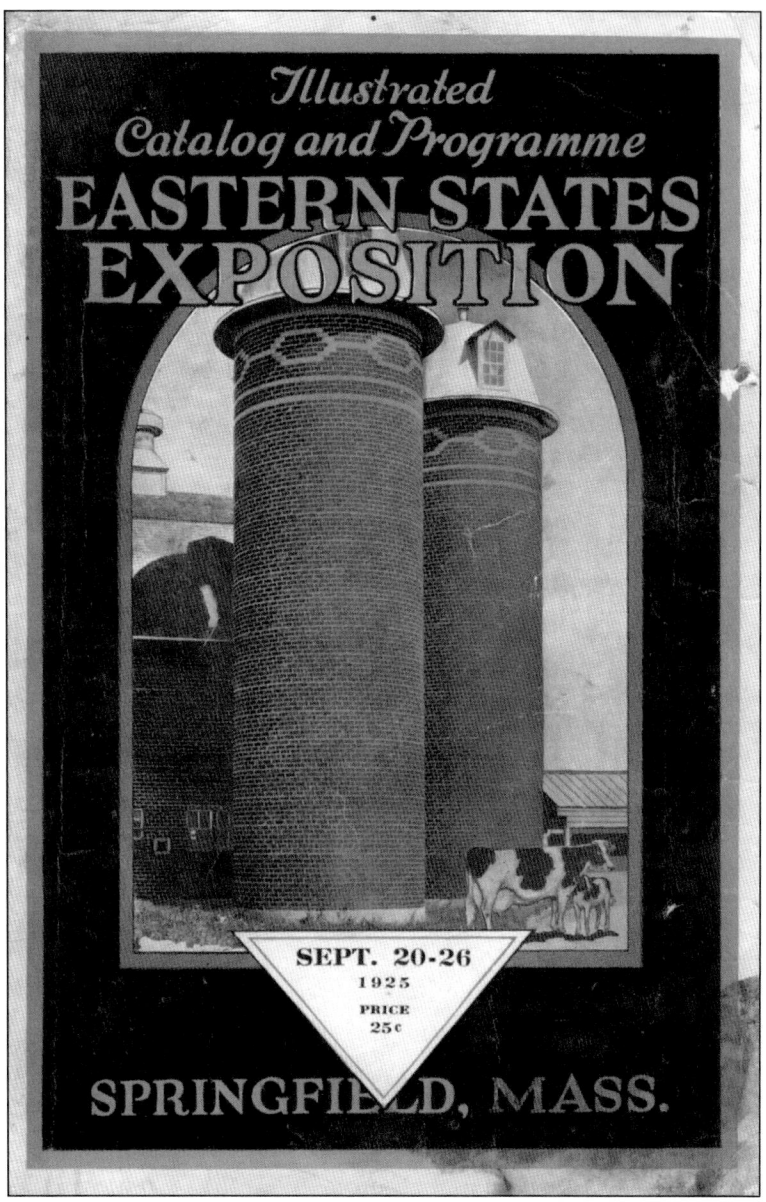

This design appeared on the cover of the exposition program for one year only, 1925. Three new buildings were opened that year; otherwise, "there will be the same type and character of exhibits as in previous years—livestock in all its classes, a poultry show, dog show, cat show, farm and industrial machinery, automobile show, state exhibits, boys' and girls' club activities, a complete junior livestock show, junior music contests, a fitter family contest, home arts, displays, fruit and vegetable shows, barnyard golf tournaments, flower show, livestock judging contests, livestock sales, etc. Bands and orchestras of national reputation will give concerts daily, four days of light harness and running racing are scheduled, there will be two days of automobile racing, the Springfield Horse Show every evening in the Coliseum arena, a complete vaudeville and circus program every afternoon and evening, and the mammoth fireworks spectacle 'Rome Under Nero' every evening at the grandstand, plus a daily display of fireworks. Throughout the Eastern States Exposition for 1925 is a quantity as well as a quality show." (Author's collection.)

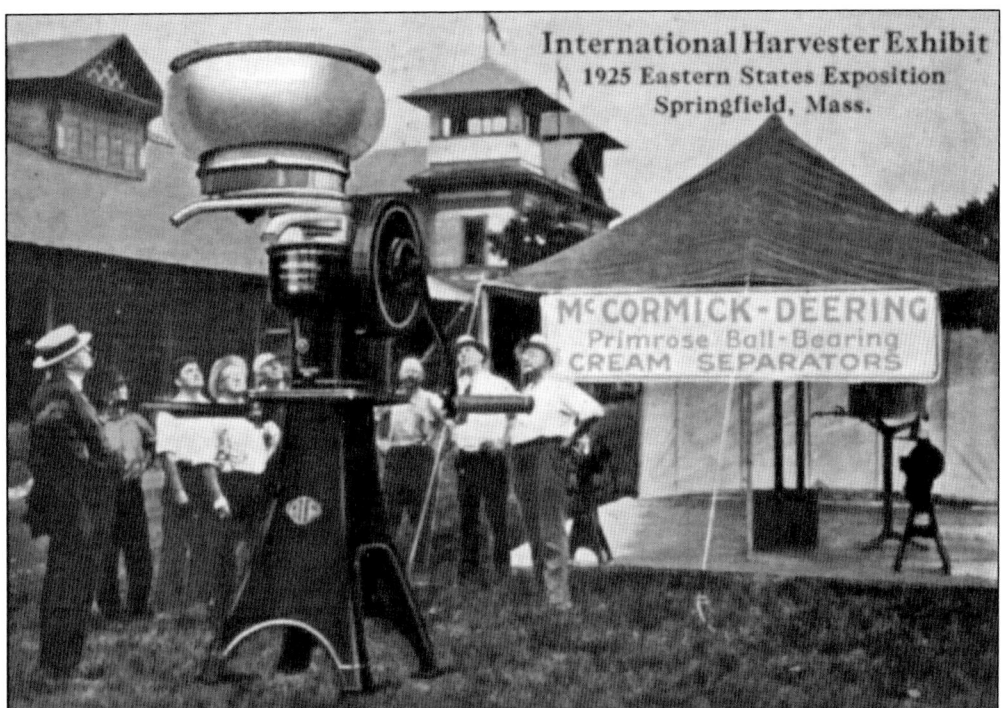

To promote its Primrose Cream Separator, the International Harvester Company utilized a giant demonstrator model that was a fixture at fairs around the country. Touting the economy and reliability of the machine, the company noted that of 200 Primroses sold in the previous decade by one dealer, only 11 had been brought in for repairs, with the largest repair bill being 89¢. (Author's collection.)

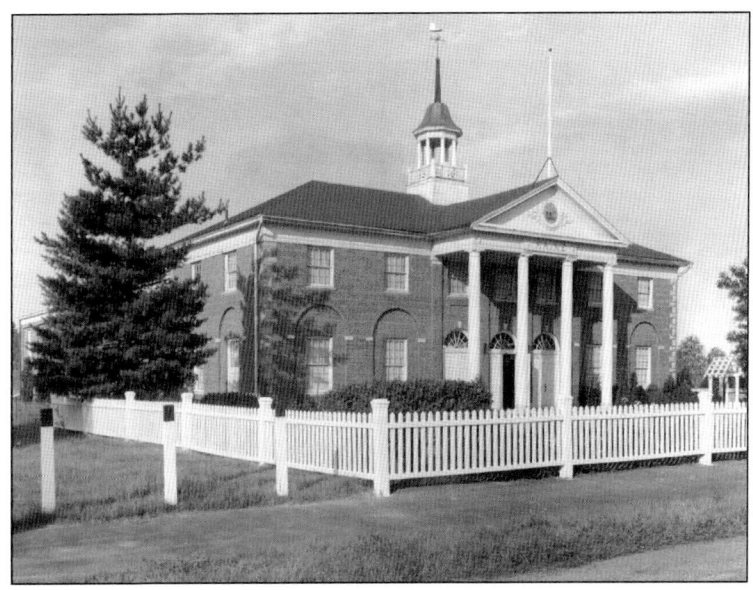

With the dedication of the $50,000 Maine Building on Tuesday, September 22, 1925, the Massachusetts Building had a neighbor and Joshua Brooks's Avenue of States became a reality. A large crowd braved lower than normal temperatures to hear Maine governor Ralph Owen Brewster and US senator Bert Fernald of Maine address those gathered.

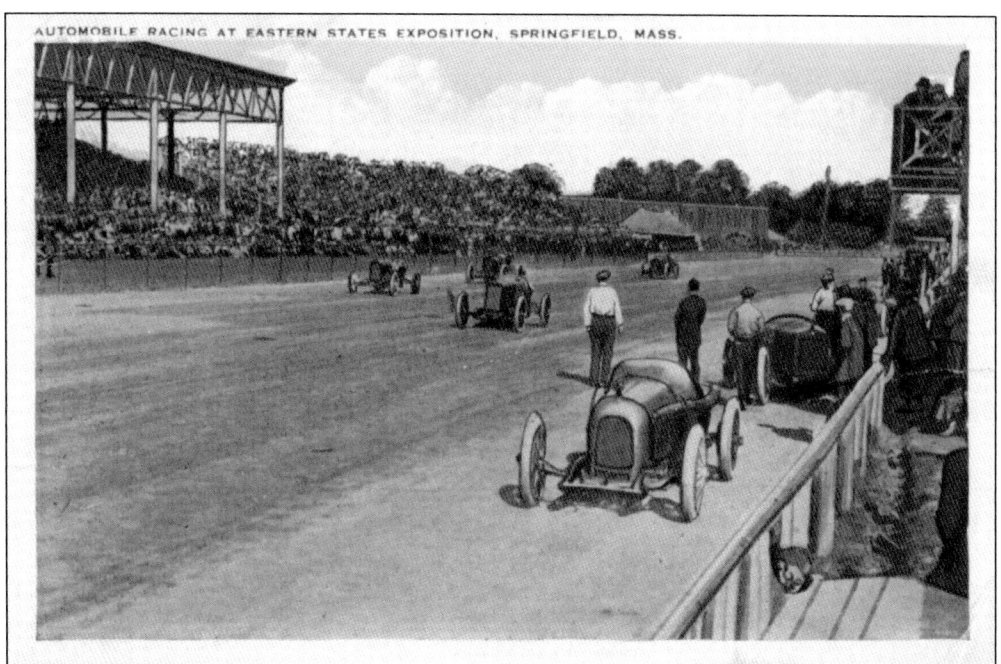

Traditional automobile racing was popular; however, during the exuberant days of the Roaring Twenties, it seems additional excitement was called for. Described in an exposition program as "the sport too fast for the movies," a game of Auto Polo "sounds like the Battle of Verdun and makes cold chills run up and down your back." Daily matches, between two Ford Model Ts "stripped to the chassis," were featured. (Author's collection.)

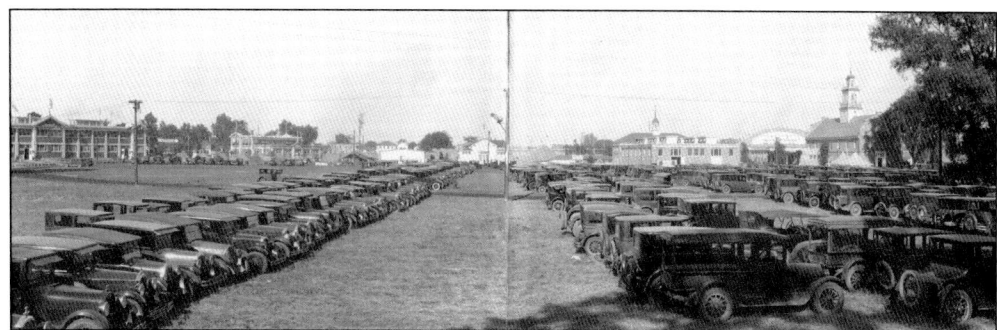

This 1926 view shows, from far left, the Junior Achievement Building, the Mohawk Village, and the Hampden County Improvement League Building. At far right is the Massachusetts Building and, to its left, the Coliseum and the Maine Building. This large grassy lot at the western end of the fairgrounds remains essentially the same after all these decades.

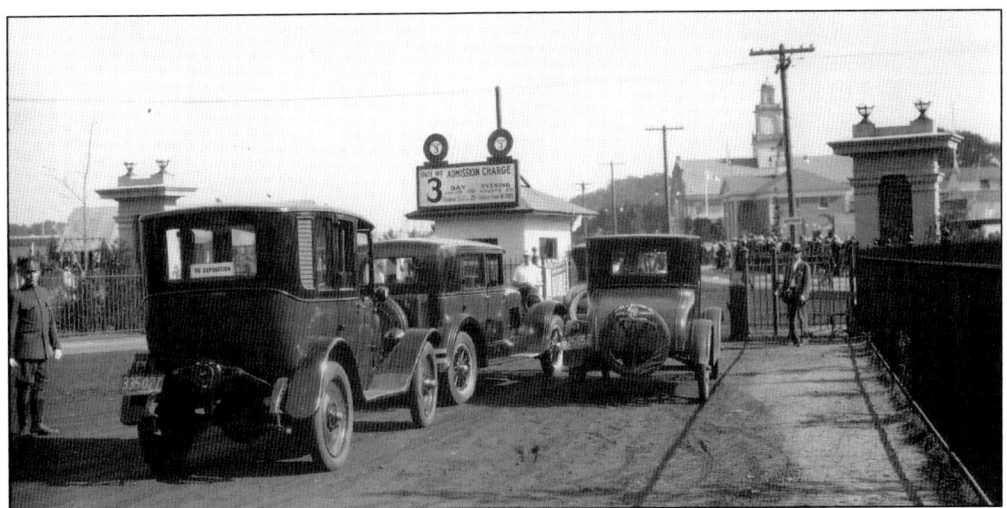

Automobiles enter Gate 3 onto the Avenue of States for the 1926 exposition. The closing day of the fair, September 25, was designated Automobile Day. As exposition promotional literature explained: "Automobile races and auto polo as well as the automobile show are high spots of the final day. Thousands of people live within easy motoring distance. New highway facilities recently completed make it an ideal place for motorists to spend their Saturday half holiday."

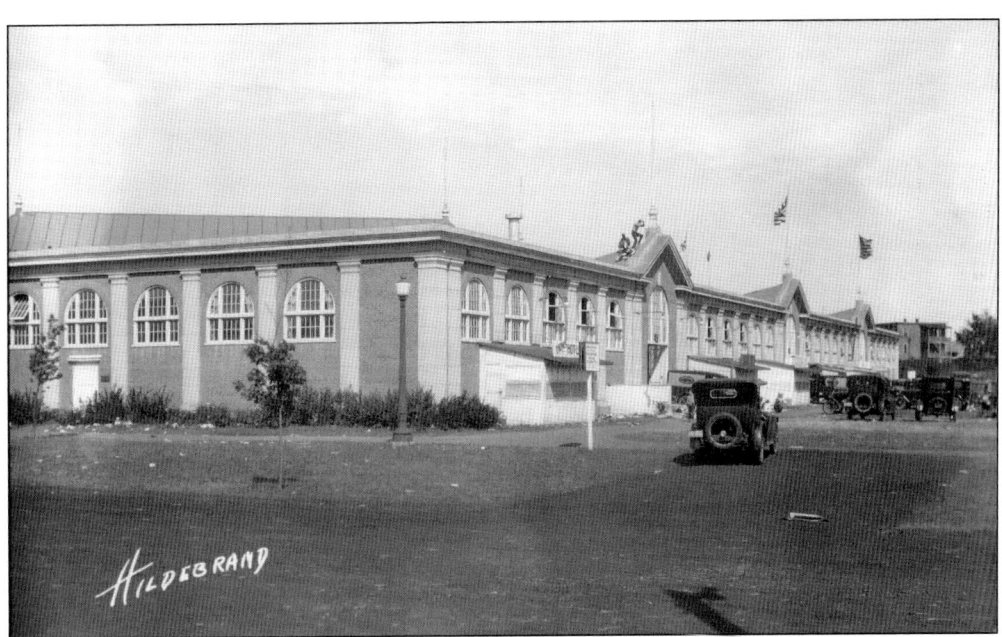

Exposition workers take down flags from atop the Industrial Arts Building at the close of the fair. The cluster of automobiles around buildings as exhibits are taken down has not changed. The arched windows and flagpoles seen here have long since been removed.

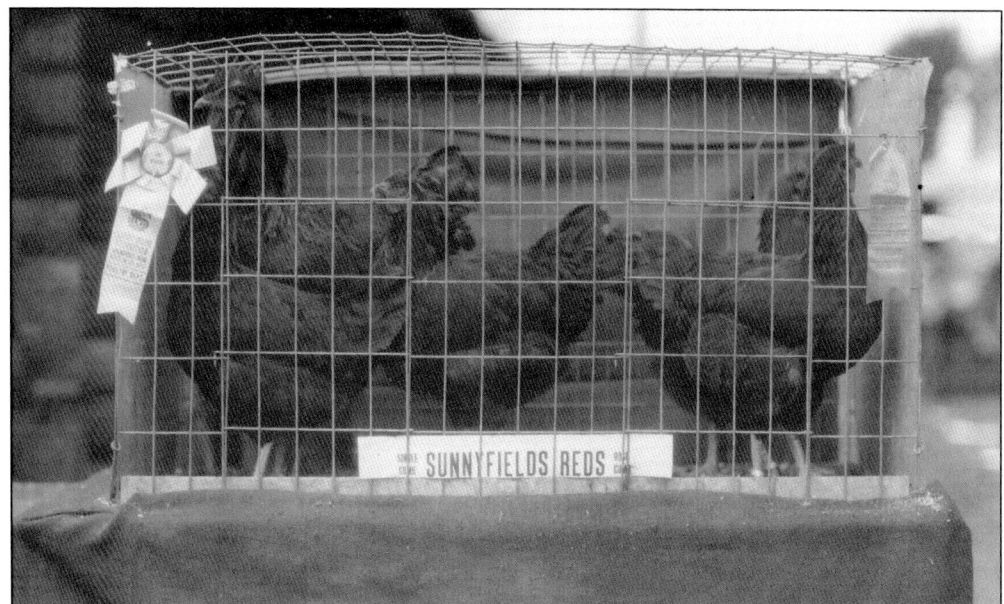

A poultry show was introduced to the exposition in 1924 and coops for 1,500 birds created in the former Machinery Hall. All standard breeds of poultry and waterfowl were included in the classifications and a special class for junior exhibitors was also added. The aim of the poultry department was to stimulate the breeding and raising of better poultry, in order to maintain and increase the reputation of the North Atlantic states for producing the finest poultry. Following the unusual success of the inaugural show, the poultry show was made an annual feature. The New England District of the American Poultry Association also gathered at the exposition in conjunction with the show.

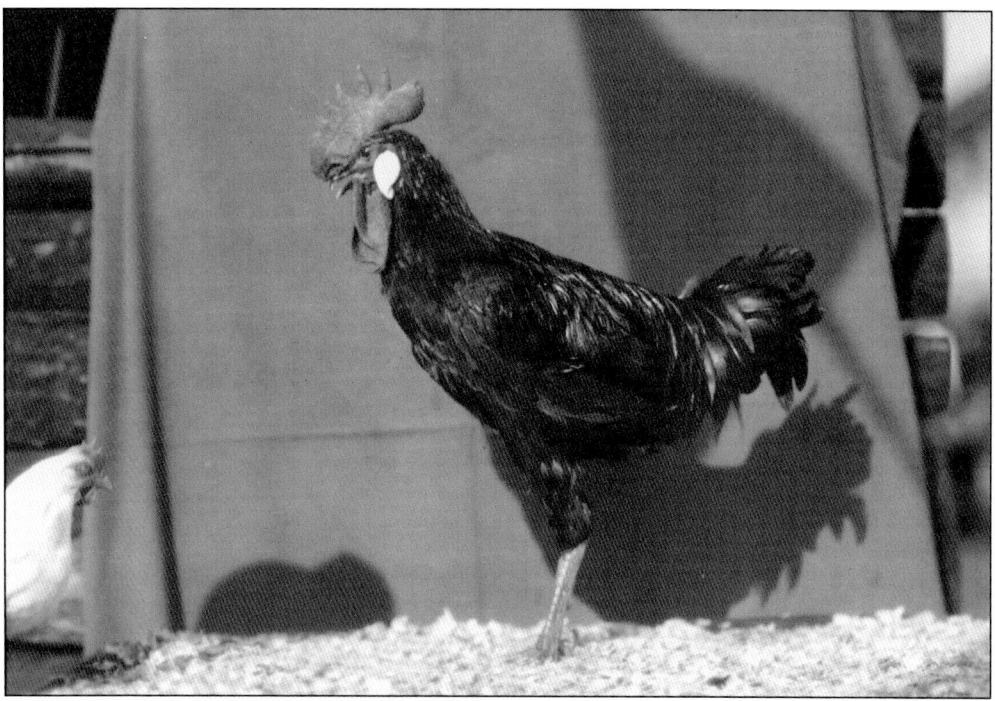

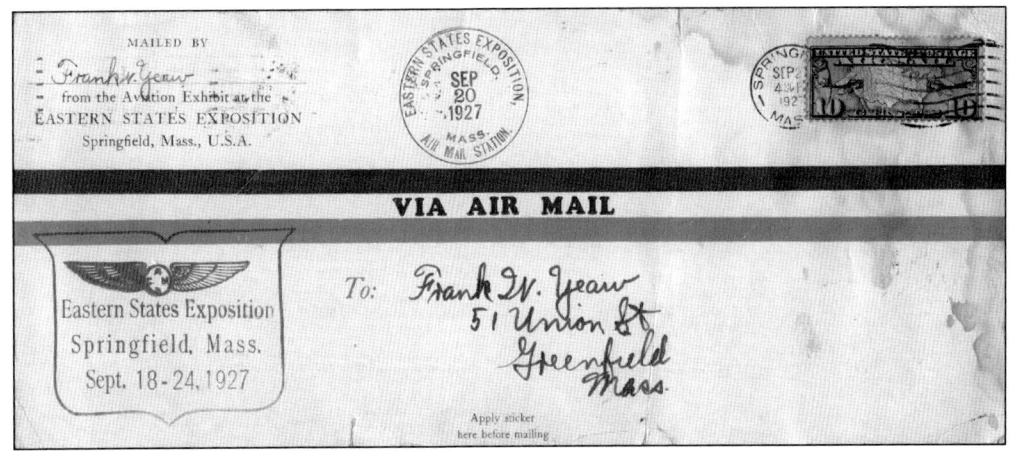

With Richard Byrd's flight to the North Pole taking place the previous year and Charles Lindbergh's transatlantic flight just four months prior, the Aviation Show at the 1927 exposition could not have been more timely. This was the first show of its kind held at any fair or exposition. In addition to the Army, Navy, and commercial airplanes on display in the Industrial Arts Building were aviation-themed exhibits by Massachusetts Airways, Colonial Air Transport, and Baush Machine Tool Company, along with one of the motors from Byrd's Fokker Tri-Motor. The souvenir airmail folder pictured above was one of 50,000 distributed at the exposition by Colonial Air Transport. Massachusetts Airways operated sightseeing flights from Randall Field in Agawam during Exposition Week; parachute jumper Walter Johnson performed there twice daily. (Both, author's collection.)

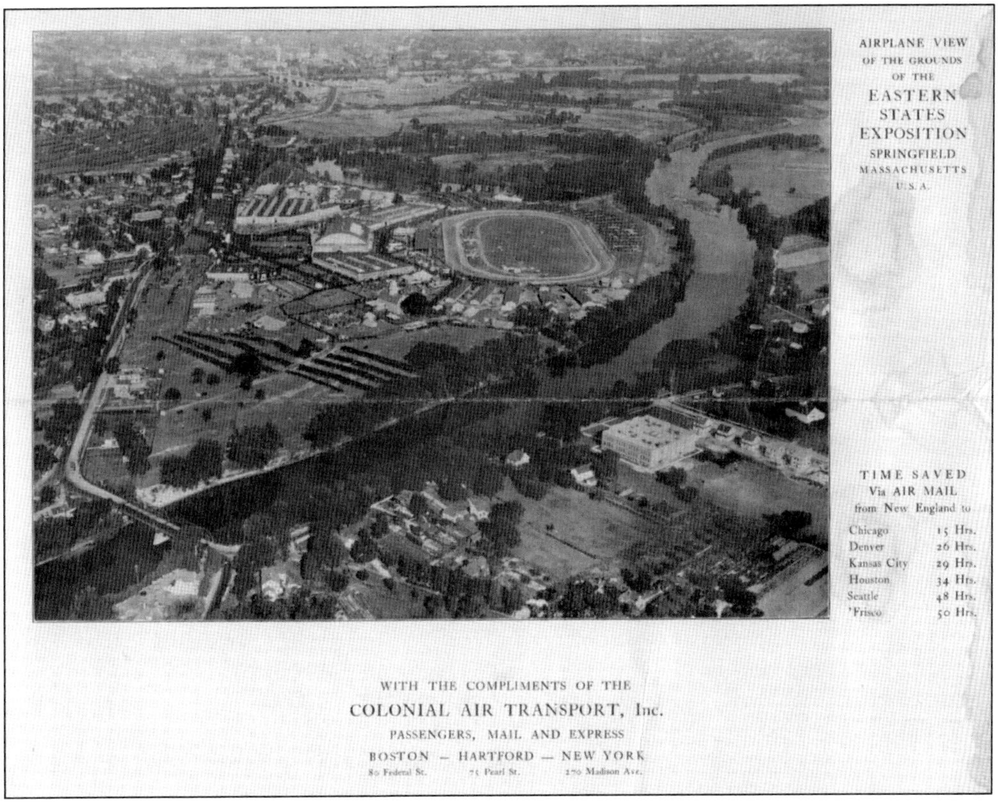

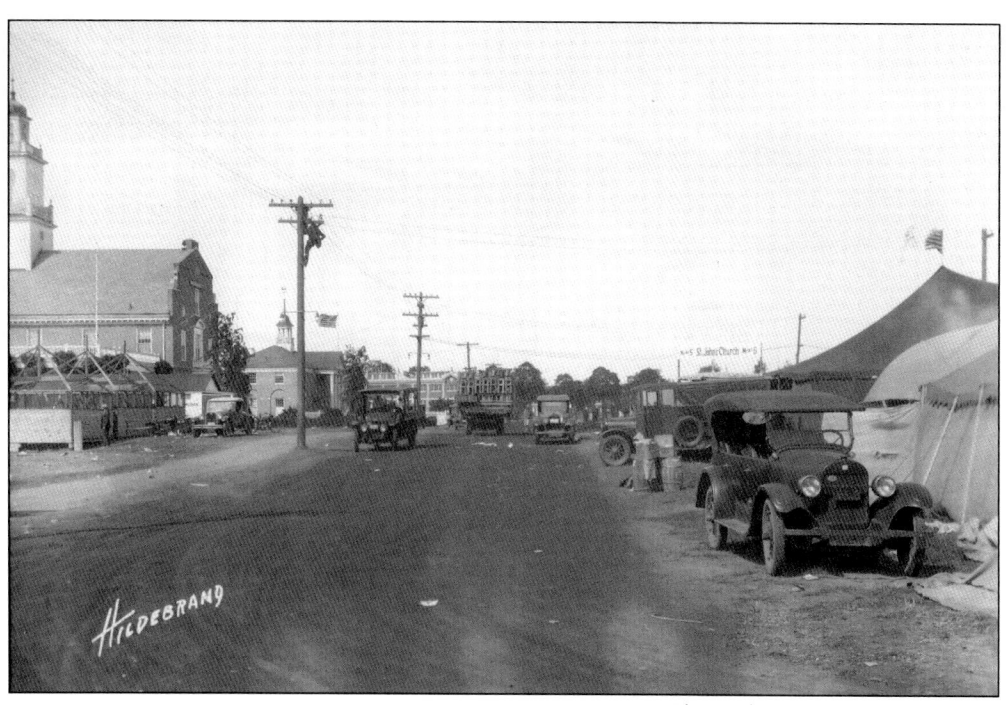

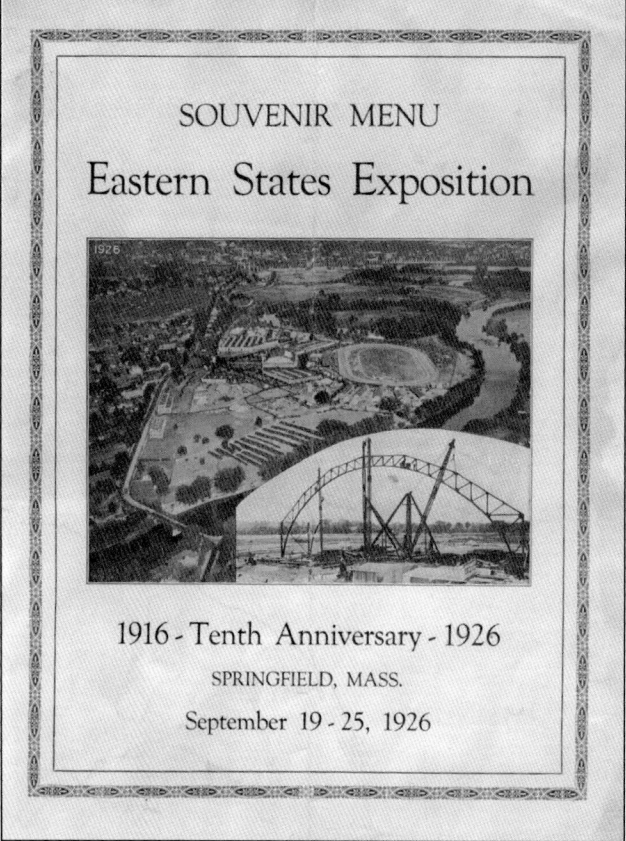

The end of the fair is always bittersweet. After an intense and exhausting week of activity, livestock is returned to the farm, tents come down, exhibits are deconstructed, fixtures hauled away, the grounds are cleaned up, and everyone returns to their normal routine for the next 51 weeks, eagerly anticipating opening day of the next edition.

To commemorate the 10th anniversary of the exposition, Springfield's Highland Hotel created a special souvenir menu. Offerings included a choice of tenderloin cutlets with creamed mushrooms and French fried potatoes for 65¢, pig's knuckles with sauerkraut and mashed potatoes for 80¢, or fresh lobster patty with asparagus tips and julienne potatoes for $1. (Author's collection.)

Two
WATER, WIND, AND WAR

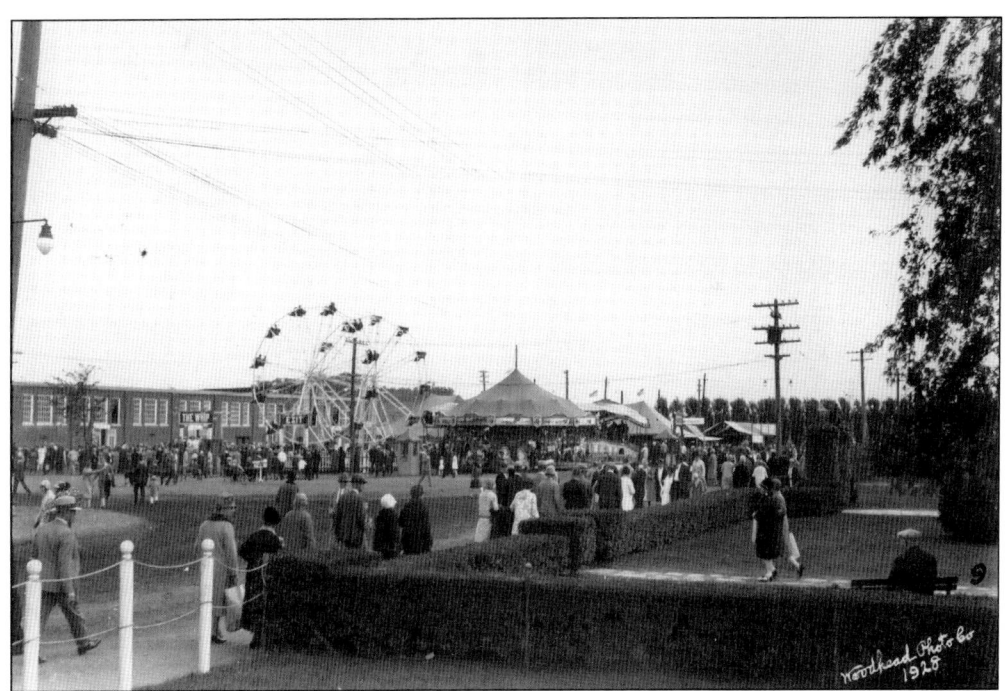

Exposition programs told of fair officials who were especially proud of the "special attractions in the form of mechanical and animal rides and playgrounds, especially for young people," declaring "Nothing is permitted to be shown or exhibited that would harm a growing boy or girl. This is a standing rule for which the Exposition is widely known and applauded for adhering to, barring all midways, sideshows, and games of chance."

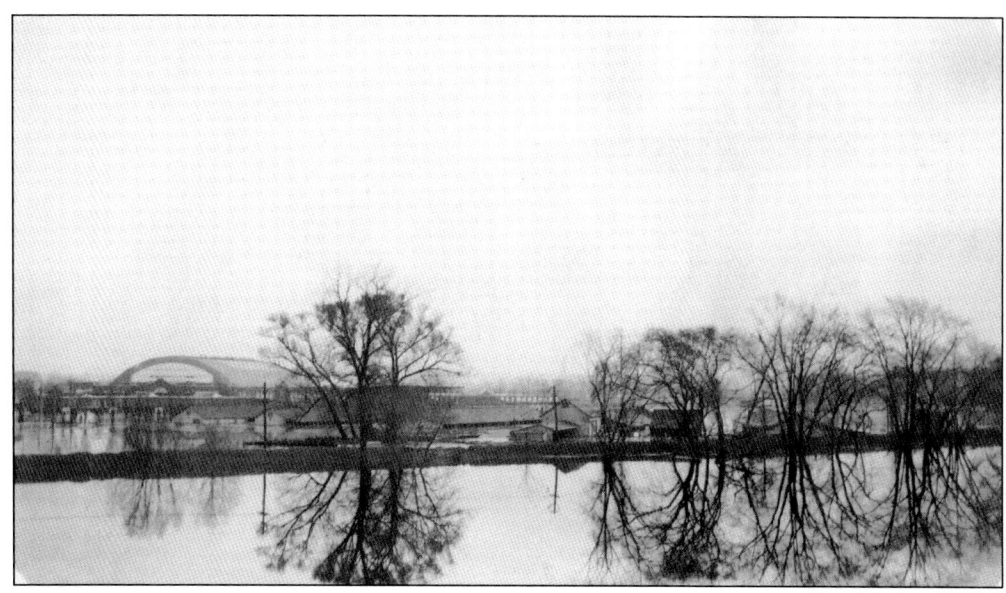

In November 1927, torrential downpours across western New England caused widespread flooding and the breach of Meadow Dike. The resulting inundation of the exposition grounds trapped 114 horses stabled there in frigid, neck-deep water. The stranded equines included 19 draft horses belonging to the Frost Trucking Company, former champion pacer Sanardo, and other show and race horses. Despite the efforts of scores of volunteers, including West Springfield and Wilson-Thompson (Agawam) Post 185 American Legionnaires, more than two dozen horses, including Sanardo, were lost, according to newspaper accounts from that time. Among the rescuers identified by the *Springfield Republican* were Roland Reed (who would become Agawam chief of police in 1950) and Ted Parsons (recipient of the Pynchon Award in 1918). Exposition president Joshua Brooks provided hot coffee and sandwiches as the men worked through the night. (Both, author's collection, courtesy of Everett Hodge.)

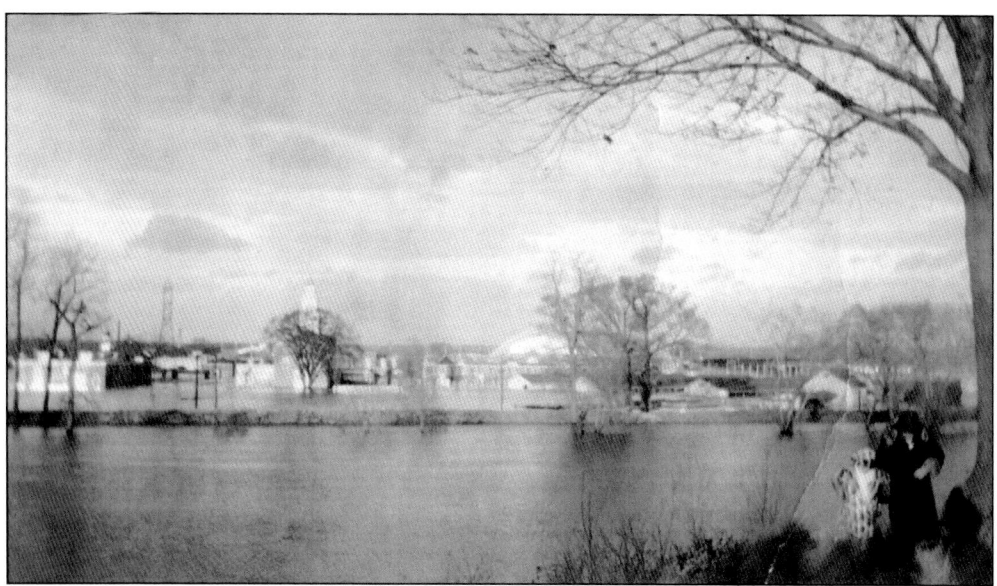

"The four fundamentals of business—production, financing, accounting, and distribution—will be in operation or depicted during the week." So was described the activities of the Junior Achievement Club within this brochure. "To every thinking man and woman interested in the future of industry and home life, the scene depicted in Junior Achievement Hall suggests a social insurance plan applicable to every community." (Author's collection.)

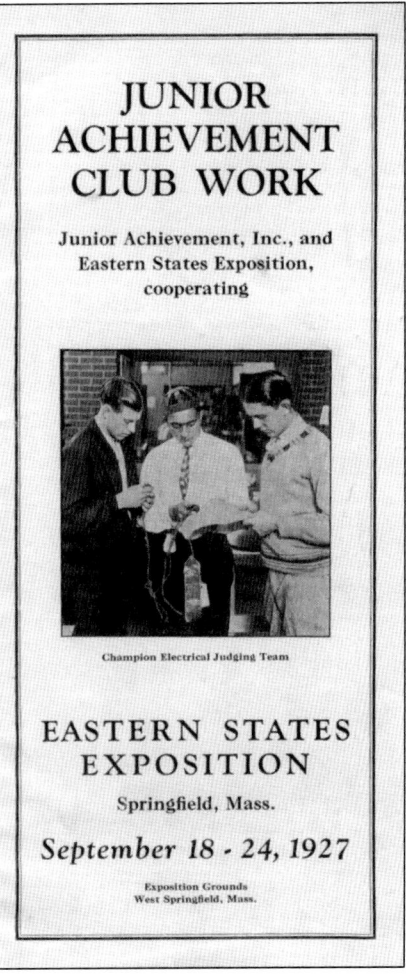

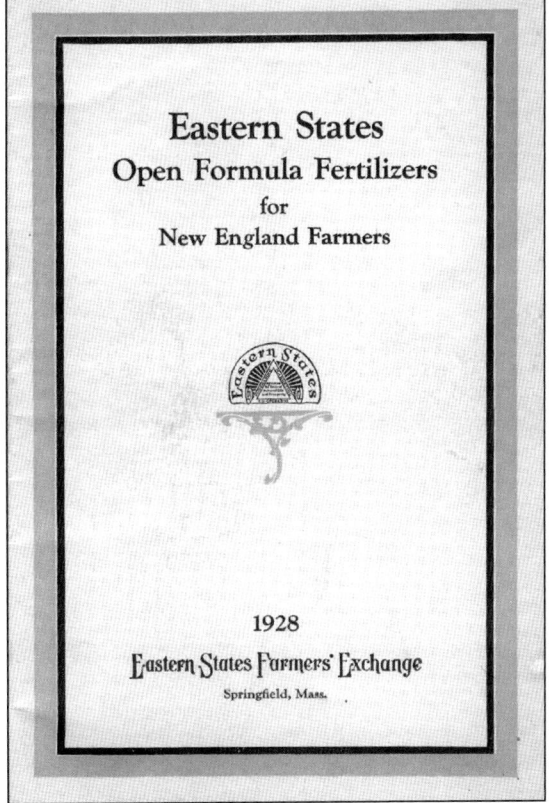

The Eastern States Farmers' Exchange was formed in 1918 "to make farming a more profitable business through methods of co-operative buying of supplies and marketing of products." Originally headquartered in Springfield, the exchange later built new offices across the river; the building now serves as West Springfield Town Hall. The Eastern States Farmers' Exchange merged with the Grange League Federation in 1964 and became known as Agway. (Author's collection.)

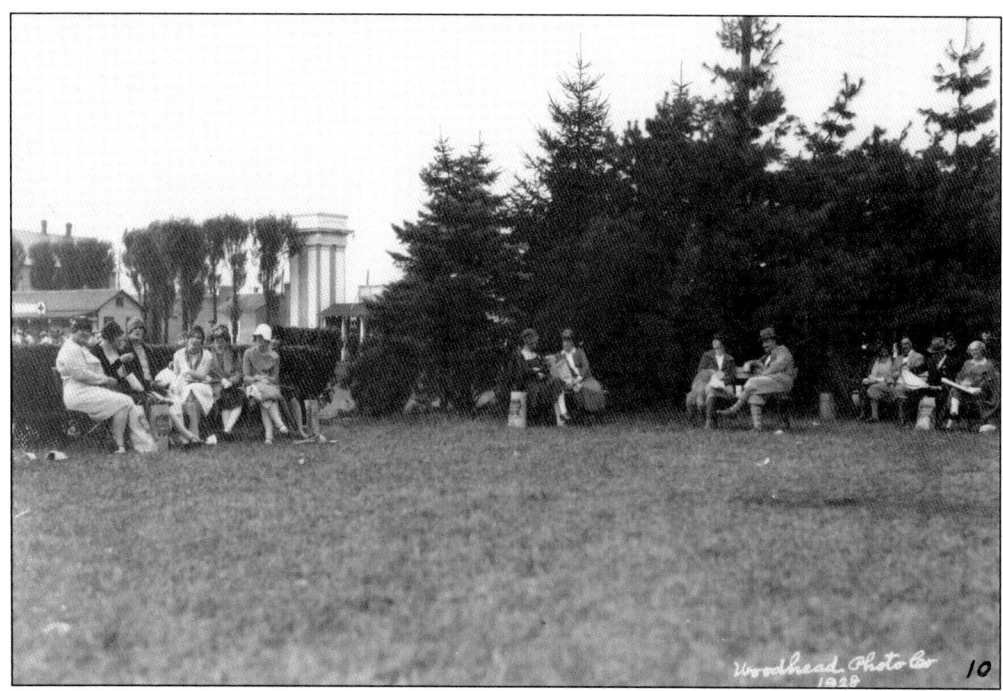

These well-dressed fairgoers from the late 1920s are taking a break near the main entrance of the exposition. Then as now, souvenir shopping bags for goods collected at the fair were much sought after. The bags here were provided by Kibbe Brothers, a popular candy maker that operated in Springfield for nearly a century (1843–1933).

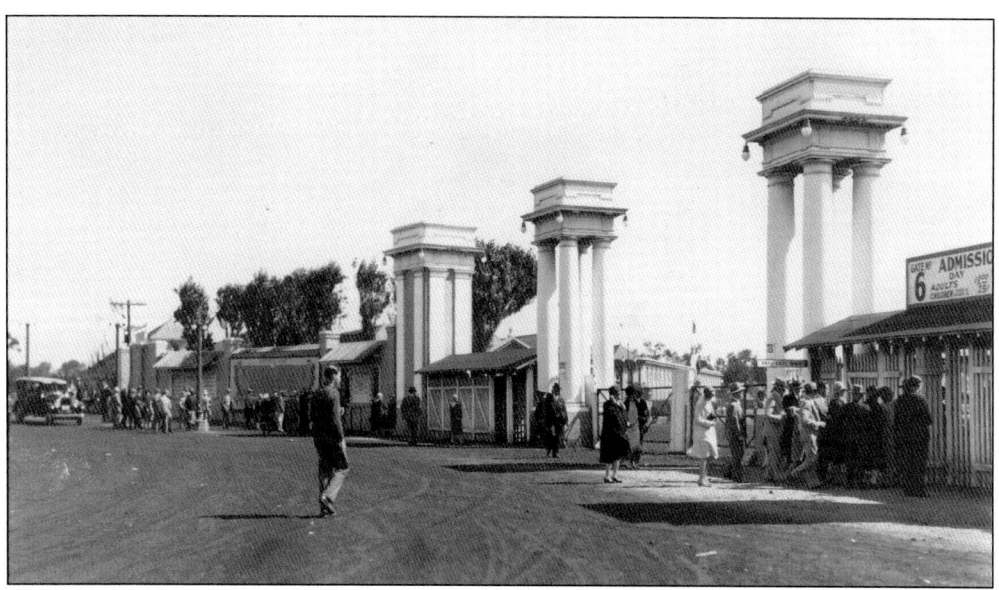

The sign at the decorative main gate advertises admission—adults $1, children 25¢. Though the ticket price has risen considerably over the years, the value remains high. The bronze tablet commemorating the George Washington Memorial Highway is found at this location, beside Memorial Avenue, in front of the Brooks Building, which replaced these pillars in 1949.

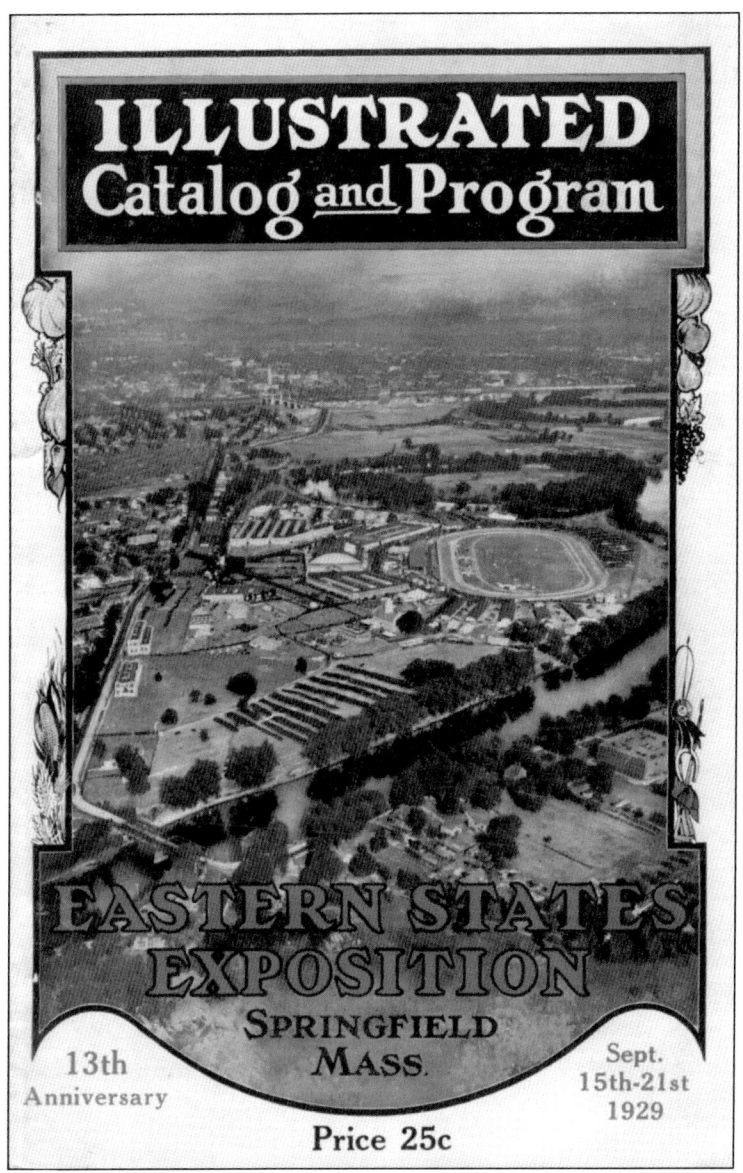

From 1926 through 1939, the exposition program featured this design and aerial view of the fairgrounds. The 1920s witnessed the Barnyard Golf (horseshoe pitching) Tournament, installation of a grounds-wide loudspeaker system (the first of its kind at any fair or exposition and, updated, still in use today), and the Dynamometer Horse Pulling Contest in cooperation with the Massachusetts Agricultural College (now the University of Massachusetts Amherst), Massachusetts Department of Agriculture, and members of the Massachusetts Agricultural Fair Association. The International 4-H Training School was introduced in 1923. In 1929, the Massachusetts Department of Agriculture installed a rock garden at the rear of the Massachusetts Building. In 1930, the exposition hosted the Massachusetts State Tercentenary Exhibit and Future Farmers of America Oratorical Contest. The following year, the New England Amateur Baseball Championships were held on the grounds; in 1932, the first Onion Show in the history of the United States and the first National Quilt Show were also held here. That year's program also described a garden exhibit "arranged for and by the unemployed men and women of Springfield and West Springfield." (Author's collection.)

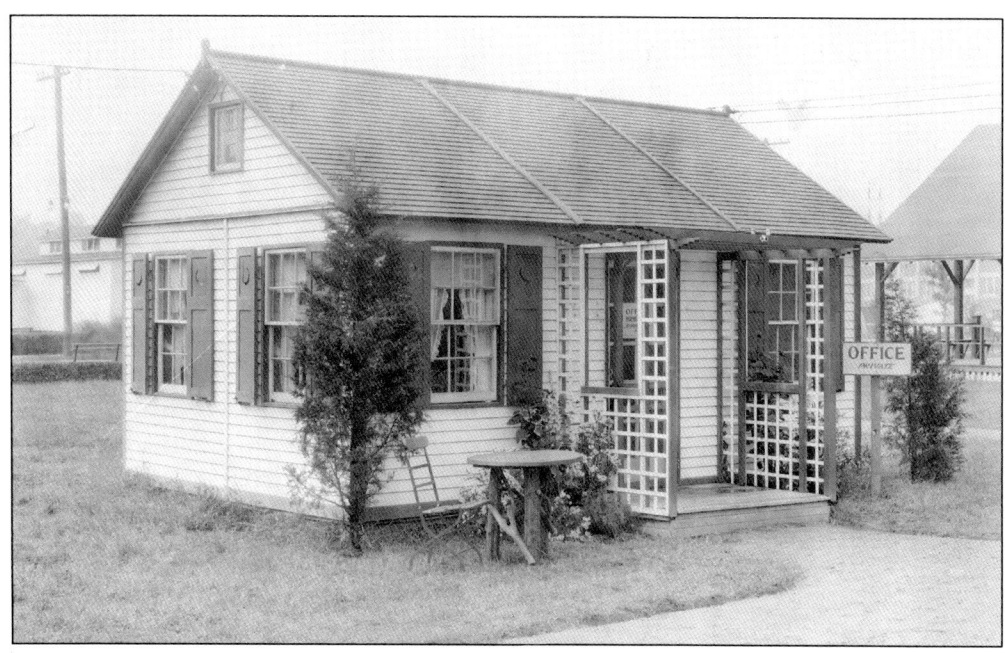

The 1920 exposition program noted that "the Home Department (under the direction of Mrs. James J. Storrow of Boston assisted by Mrs. Schuyler F. Heron of Boston and Mrs. A.C. Dutton of Springfield) has a place of its own this year on the Exposition grounds for the first time." Fourteen separate organizations cooperated to present lectures, demonstrations, and exhibits of work by women in five cottages surrounded by shrubbery, flowers, lawns, and a model kitchen garden. According to Dennis Picard, director of the exposition's Storrowton Village Museum, the temporary cotages were "not-so-affectionately referred to as "dog houses." Katherine Heron, whose office is seen above in 1927, was a member of the committee that first advocated replacing the cottages with what would initially be known as the New England Colonial Village. The department's Home Information Center is seen below.

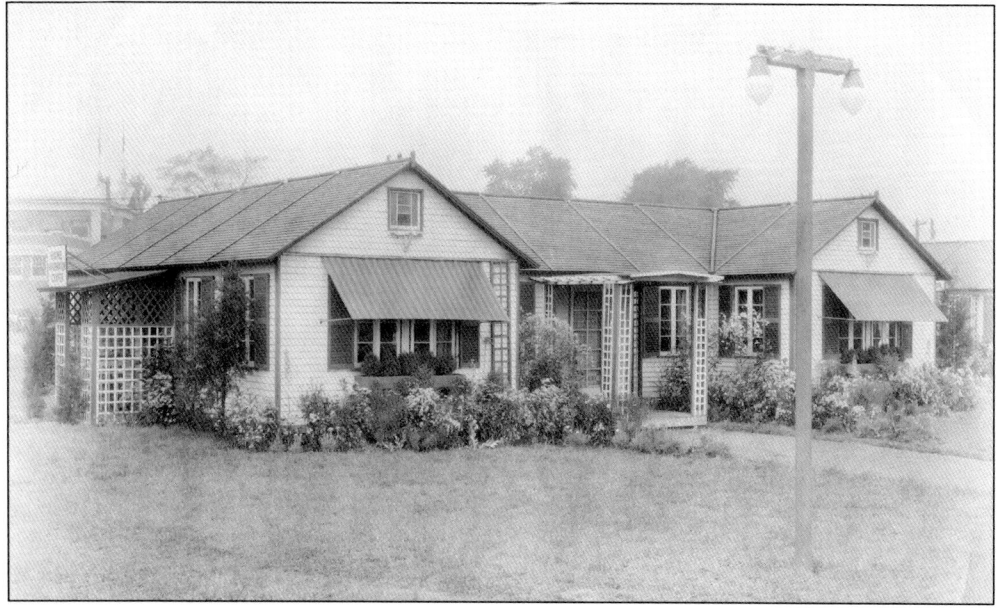

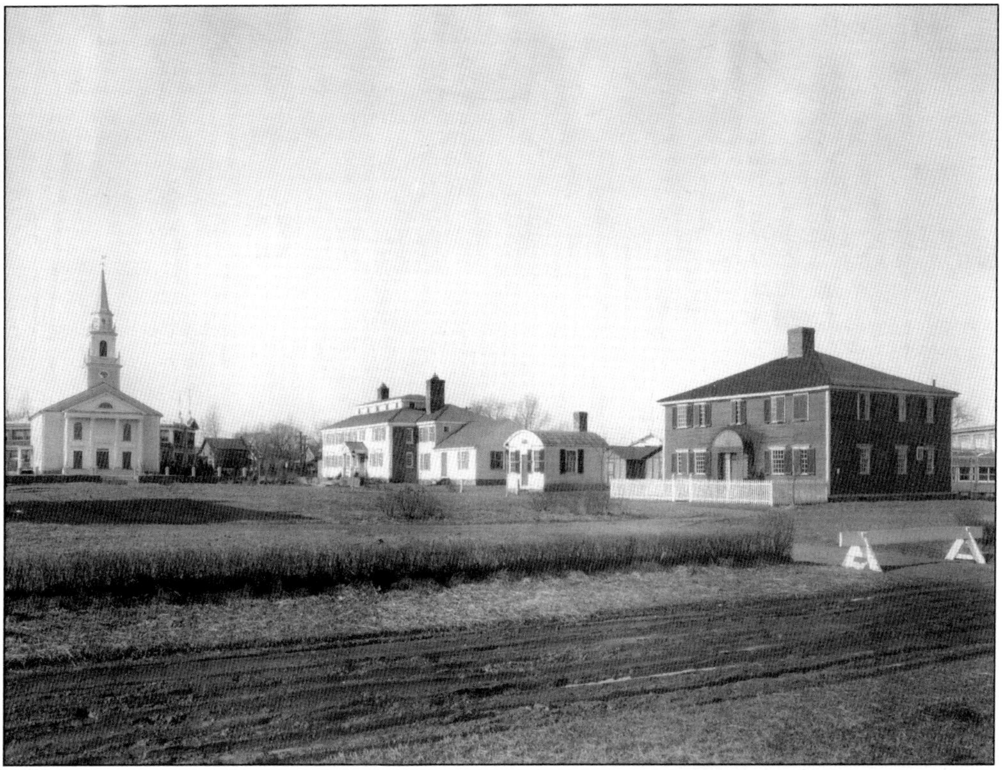

The Home Department took the "first step in the process of changing from temporary to permanent buildings" when the 1794 Gilbert homestead, at far right above, was moved from West Brookfield, Massachusetts, to the exposition grounds in 1927 through the generosity of benefactress Helen Storrow. Joining it in 1929 were, continuing to the left, the 1810 Eddy Law Office, from Middleboro, Massachusetts; the 1776 Potter Mansion, from Brookfield, Massachusetts; and the 1834 Union Meeting House, from Salisbury, New Hampshire, also pictured at right being reconstructed. The 1789 Atkinson Tavern had already been relocated from the Quabbin Reservoir–doomed town of Prescott, Massachusetts, in 1928. The 1767 Phillips House, from Taunton, Massachusetts; the 1810 Little Red Schoolhouse from Whately, Massachusetts; the 1850 Chesterfield, New Hampshire, blacksmith shop; and the 1822 Southwick (Massachusetts) Baptist Meeting House were all added in 1930, completing the New England Colonial Village.

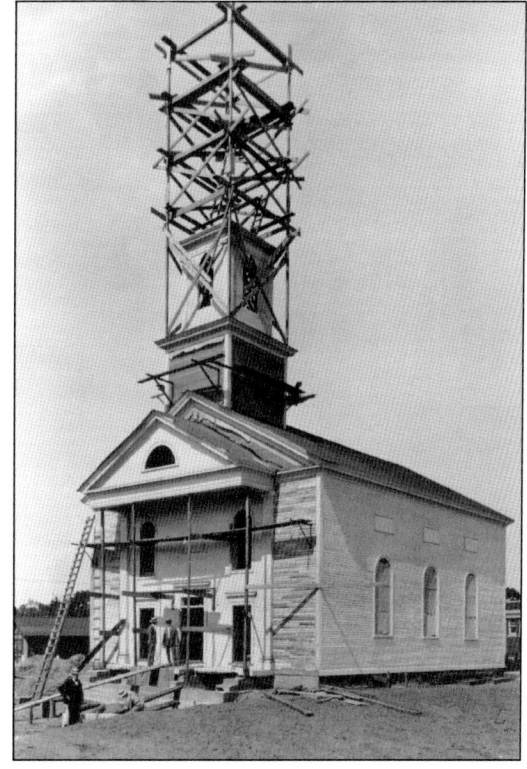

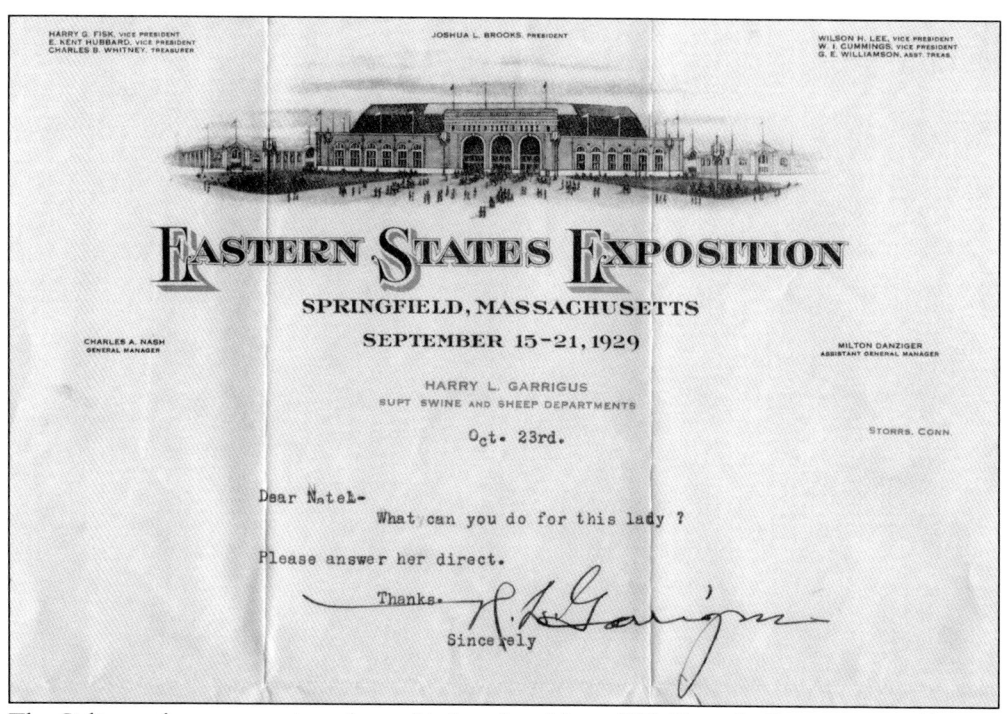

The Coliseum features prominently on this letterhead from 1929. Harry L. Garrigus, noted here as superintendent of the swine and sheep departments, was for many years chairman of the animal husbandry department at the University of Connecticut. To this day, the exposition and UConn enjoy a close relationship. In 1939, the school's Block and Bridle Club organized a Little Eastern States Exposition, modeled after the West Springfield fair, on the Storrs campus. (Author's collection.)

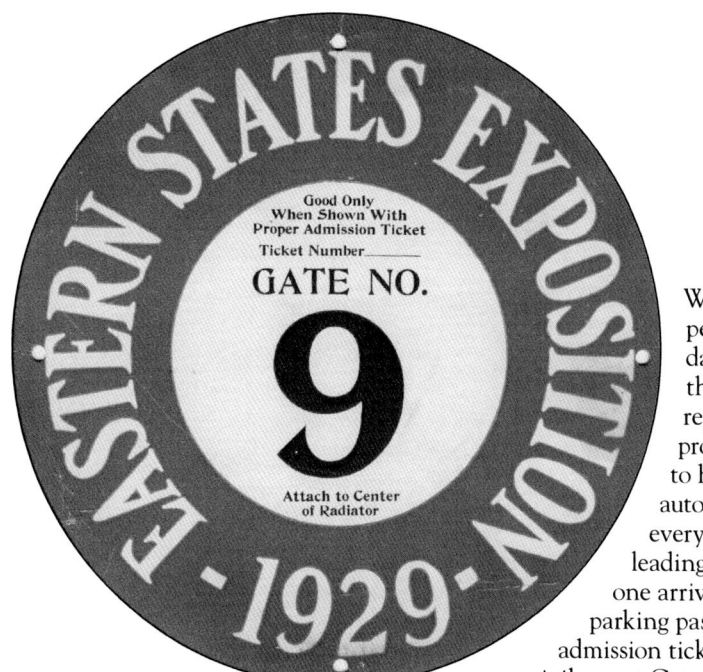

When a record 60,000 people attended opening day of the 1929 exposition, the *Springfield Republican* reported that "traffic lanes proved entirely inadequate to handle the throng of automobiles that jammed every highway and side road leading to the grounds." Once one arrived, this cardboard parking pass, in combination with an admission ticket, would allow parking privileges at Gate 9. (Author's collection.)

Construction of the Vermont State Building began in June 1929. With artisans putting the finishing touches on the building as of September 10, it was completed just in time for opening day. Former president Calvin Coolidge, a native of Plymouth Notch, Vermont, and his wife, Grace, attended the September 17 dedication ceremony, which was held indoors on account of rain.

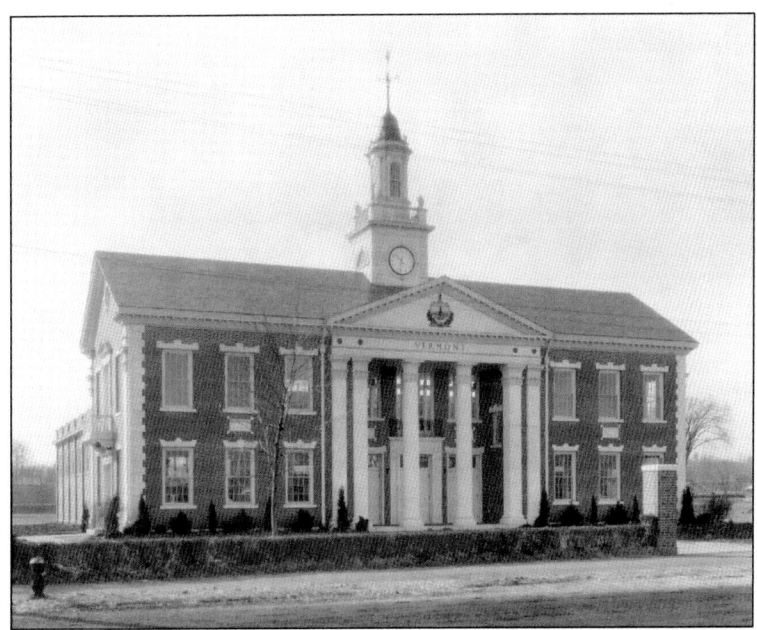

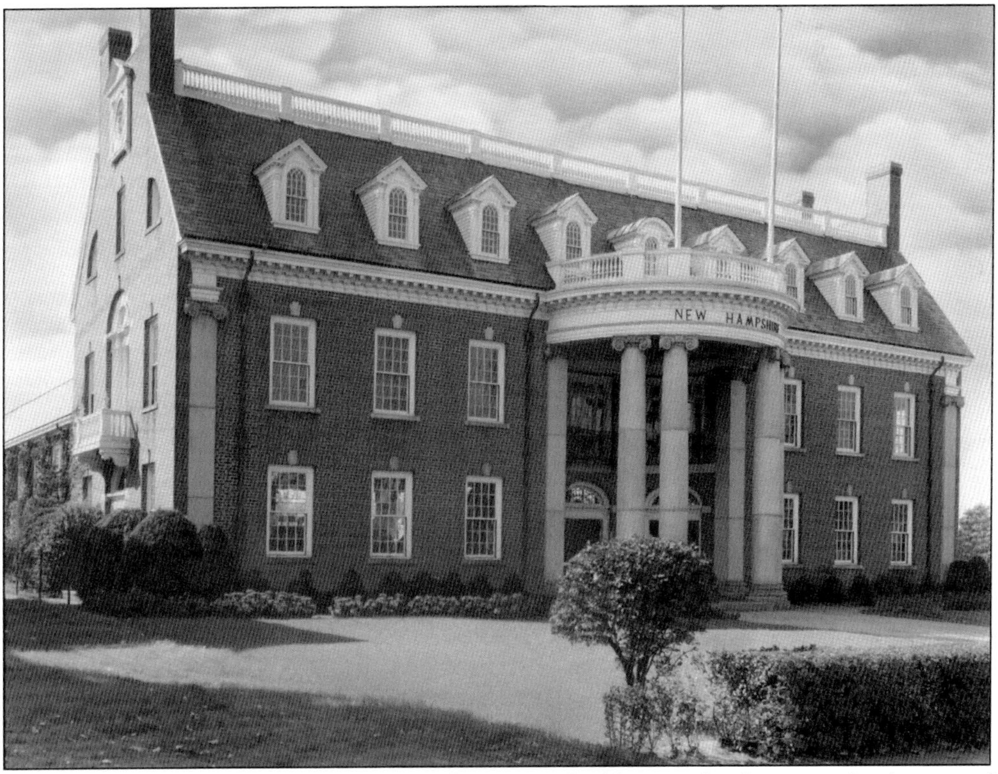

The local press heralded the New Hampshire State Building as the finest on the Avenue of States when it was completed in 1930. Unfortunately, opening day visitors were "obliged to use improvised wooden walks" to enter and exit the building because the "colored map of the state which forms the main walk was not yet quite dry." Among the building's exhibits was a cross section of a 320-year-old tree.

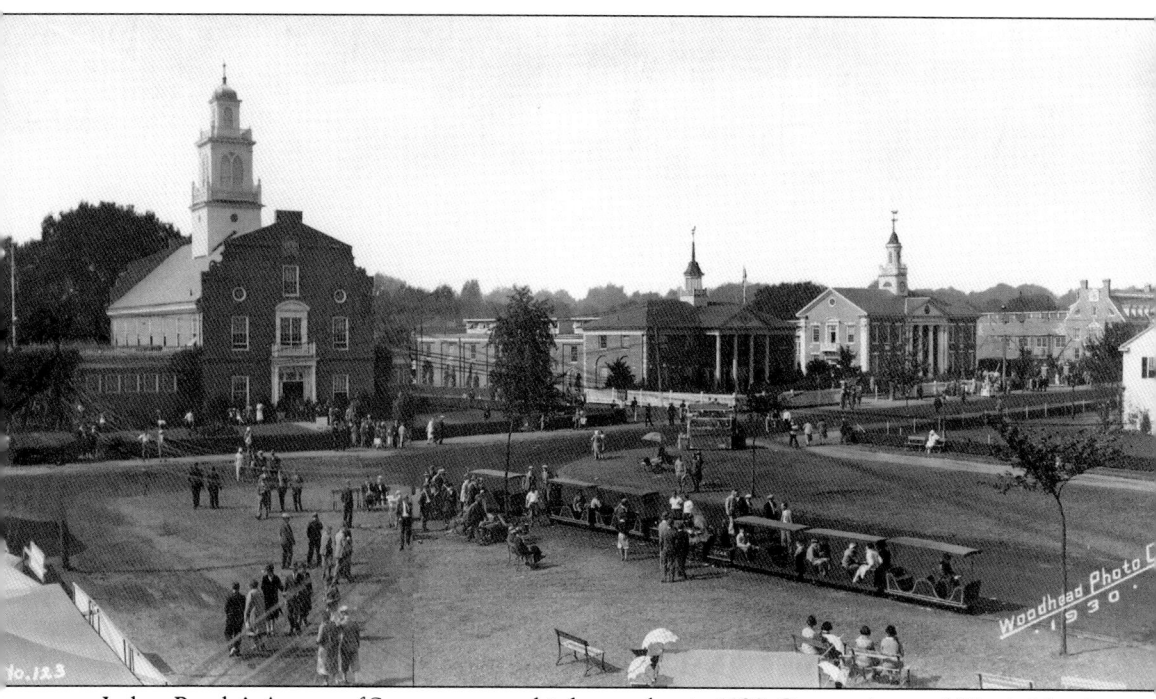

Joshua Brooks's Avenue of States was two-thirds complete in 1930; Connecticut and Rhode Island had yet to commit to his grand vision. From left to right, the Massachusetts, Maine, Vermont, and New Hampshire buildings proudly represent their states with exhibits devoted to the promotion of their agricultural, industrial, commercial, educational, natural, and recreational resources. During Exposition Week, the *Springfield Republican* carried dozens of fair-related dispatches, among them notice that aviator Lowell Bayles had performed stunts in his Gee Bee Sportser in the skies above the fairgrounds, and a report of that year's cat show: "The oppressive heat which made yesterday so uncomfortable was felt keenly within the cat show tent. Several of the felines had convulsions in the morning, and this gave their owners fits, figuratively speaking." Also mentioned was the "appropriate name for the cat show's only judge: Miss Ethel R.B. Champion." In 1930, area schools were closed so local pupils could attend on Monday, Children's Day, a practice that ended in 1955. These days, tickets issued to students are valid only after 2:00 p.m.

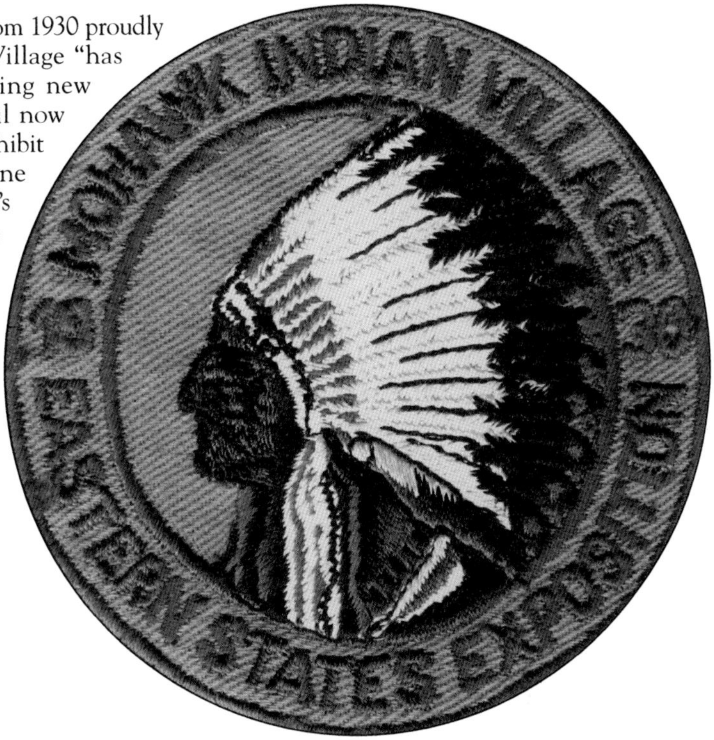

The exposition program from 1930 proudly stated that its Mohawk Village "has grown year by year, adding new projects and features until now it is the foremost Scout exhibit in the United States." One component of the exposition's Scout exhibit included a large 25-booth display of the merit badge program. (Author's collection.)

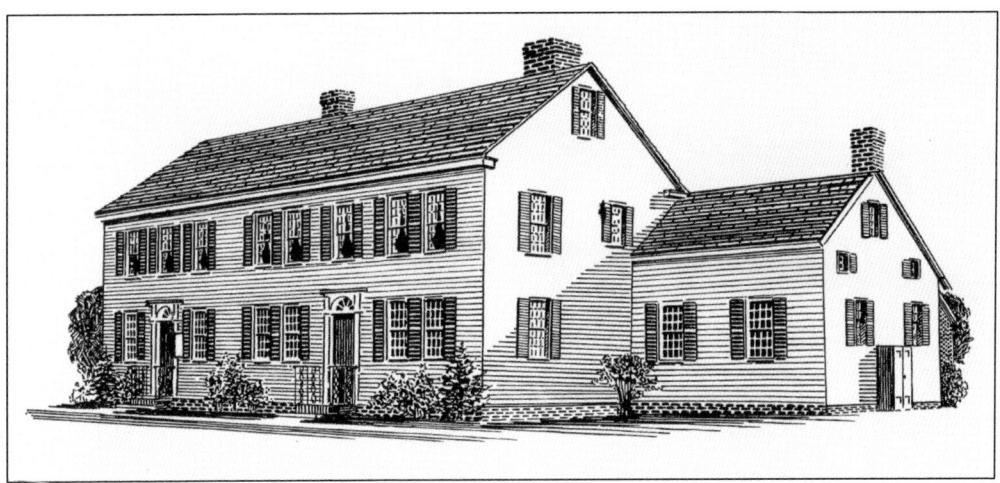

This illustration of the Atkinson Tavern was used in the 1930 brochure announcing the completion of Helen Storrow's New England Colonial Village. In 1944, the village was renamed Storrowton Village in her honor. In 1947, the second floor of the tavern was outfitted as a radio center with a comfortable lounge, available as a location for interviews and special commentator broadcasts. (Author's collection.)

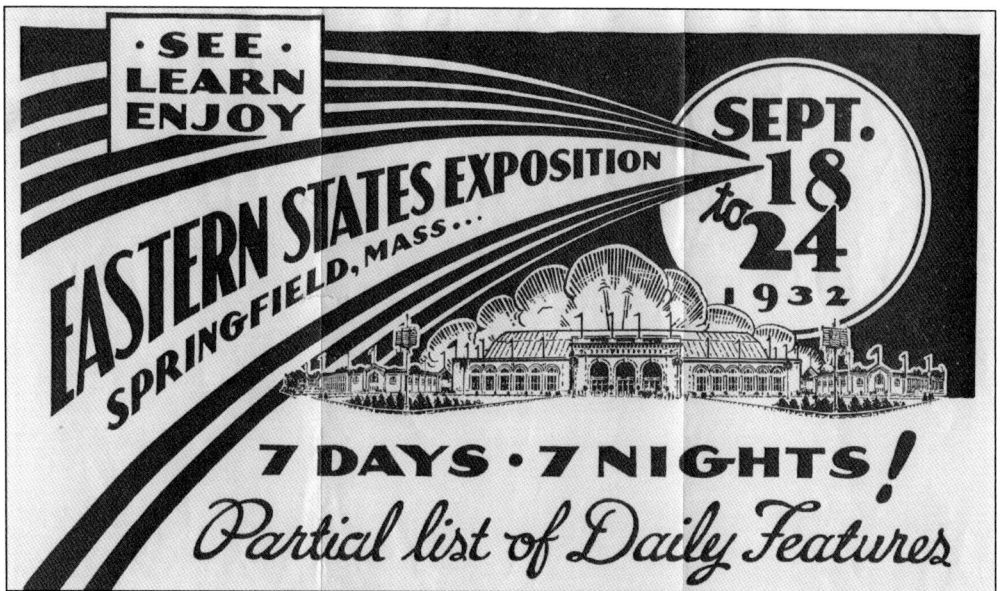

"See. Learn. Enjoy." Education has always been an important component of the exposition. For a century, the latest agricultural innovations and best practices have been promoted to farmers, and homemakers have been introduced to labor-saving devices. The exposition's youth programs have garnered national acclaim and served as models for countless others. (Author's collection.)

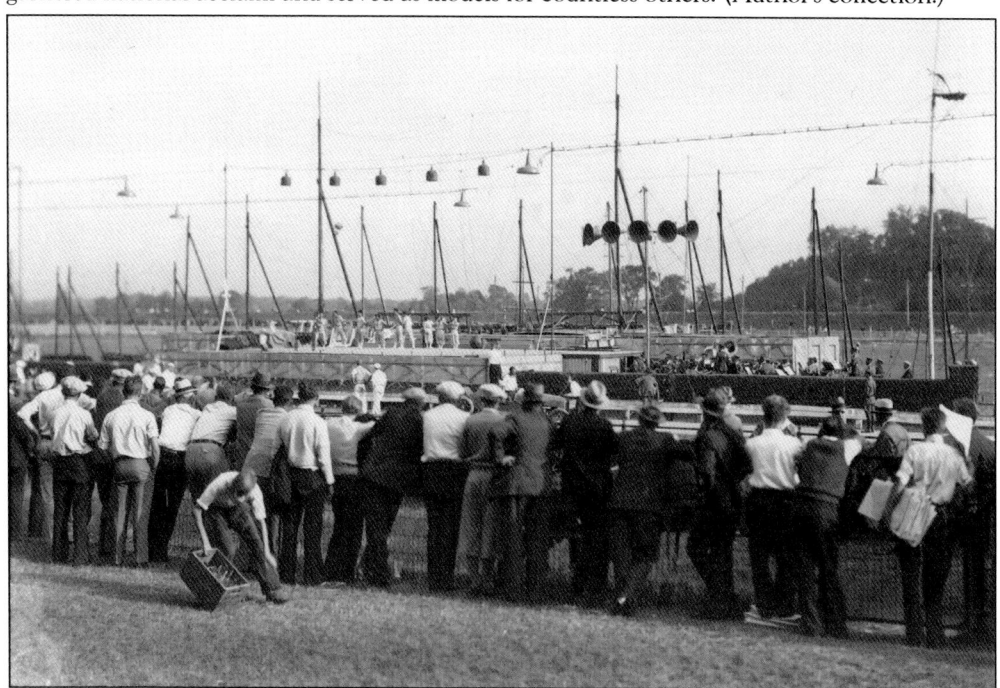

The Grandstand Night Spectacle offered each evening during the 1932 exposition included "leading acts from the best circuses and vaudeville circuits of America and Europe," according to that year's program. Seen here on stage is the Wan Wan San Troupe, a group of "sensational Oriental performers who have brought all of the mysticism of acrobatic and juggling feats known to the Far East." At right is the 104th Infantry Band, which gave performances each afternoon and evening.

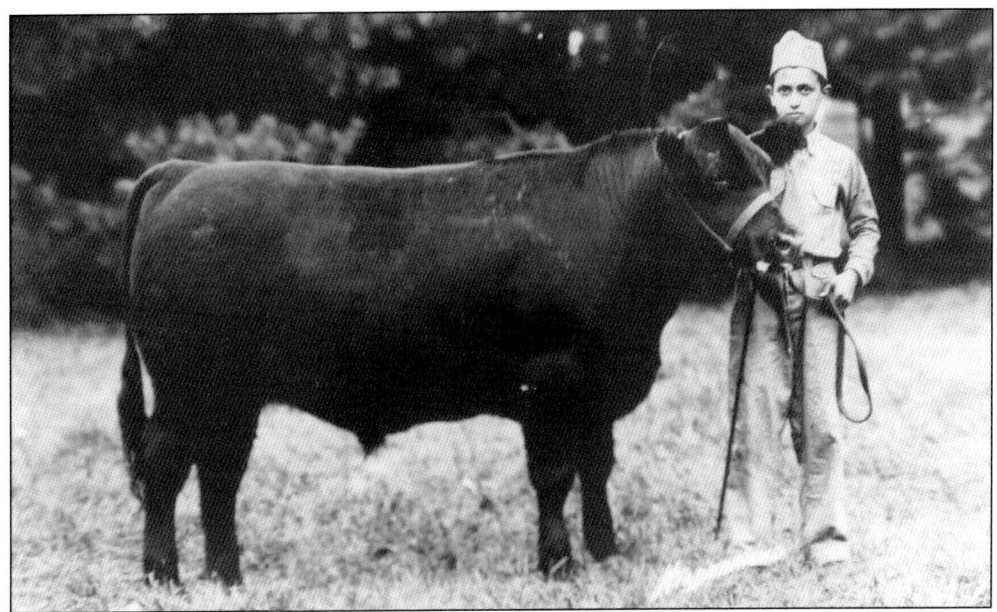

Of the 1932 4-H Baby Beef contest, the *Springfield Republican* claimed "Feeding Hills (Massachusetts) Triumphs. The Aberdeen Angus heavyweight division produced a very acceptable leader in Gold, shown by Albert Christopher of Feeding Hills, but young Lee T. Jenks of the same town with Briarcliff Rocket was right at his heels, and yet another Feeding Hills youngster, Sabatina DePalma was third." Christopher, seen here, would later be better known as Agawam's "Corn King." (Courtesy of Bill Fearn.)

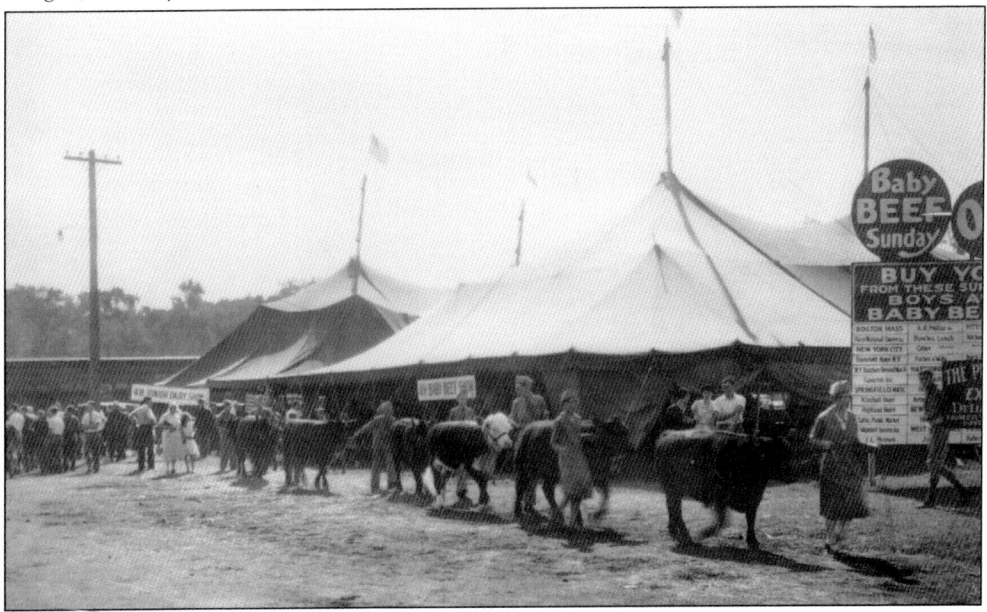

More than 85 baby beeves were raised in 1932. That year's grand champion, Stub T, was raised by Harold P. Hamilton of Pine Plains, New York, and sold to the Great Atlantic & Pacific Tea Company (A&P) for a record $2.10 per pound. Baby Beef judge W.J. Kennedy felt that the Aberdeen Angus steer was the "best for its age that ever entered an American show ring." (Courtesy of Bill Fearn.)

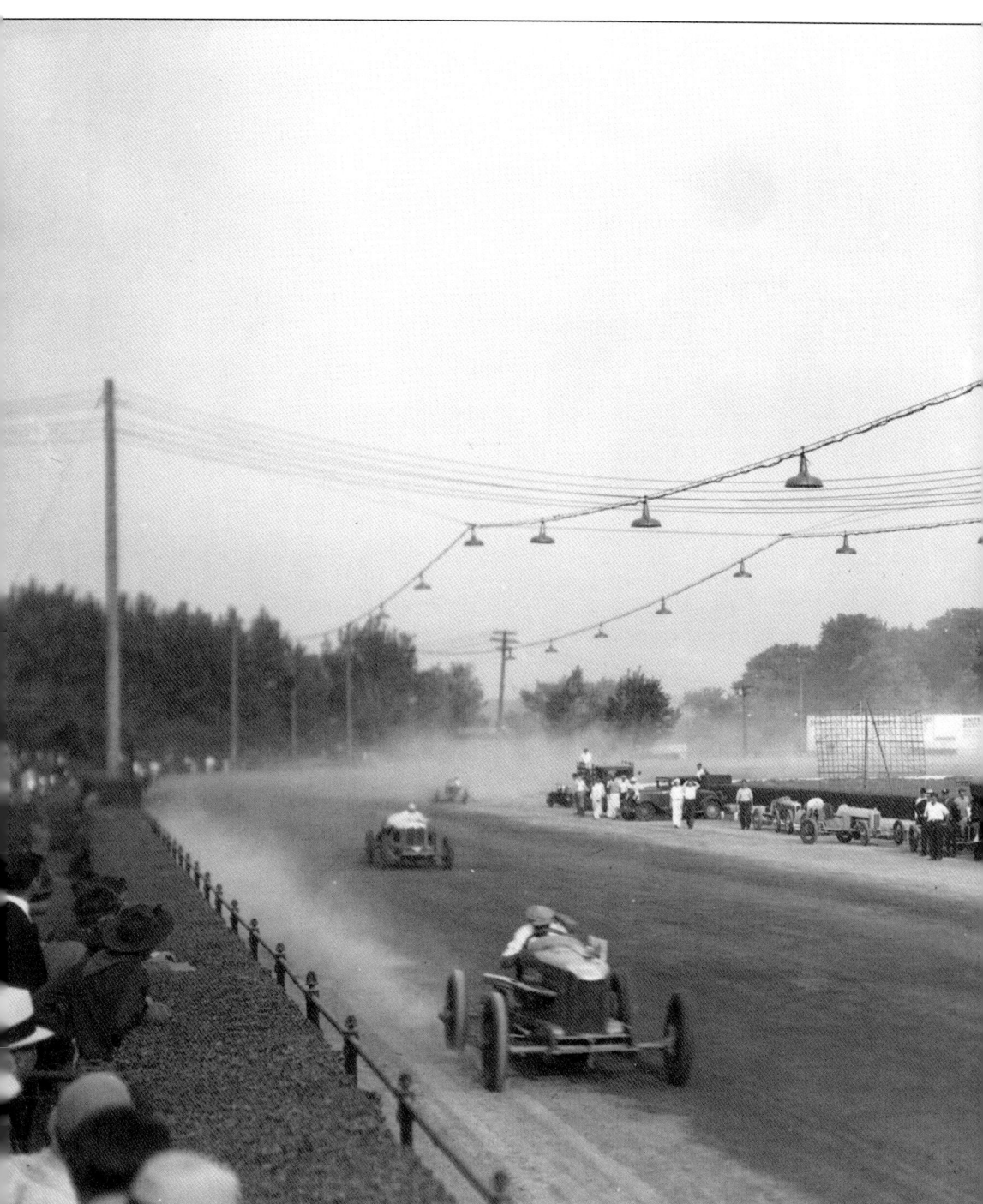

On the last day of the 1932 exposition, more than 10,000 spectators witnessed Sig Haugdahl win the New England Sweepstakes automobile race in his Miller Eight racer and Swan Peterson take the Bay State Derby event in his Duesenberg. The previous day, Haugdahl won the Eastern States Sweepstakes. Haugdahl was International Motor Contest Association champion for six years running, from 1927 to 1932. Speed enthusiasts could also view the Thompson Trophy–winning

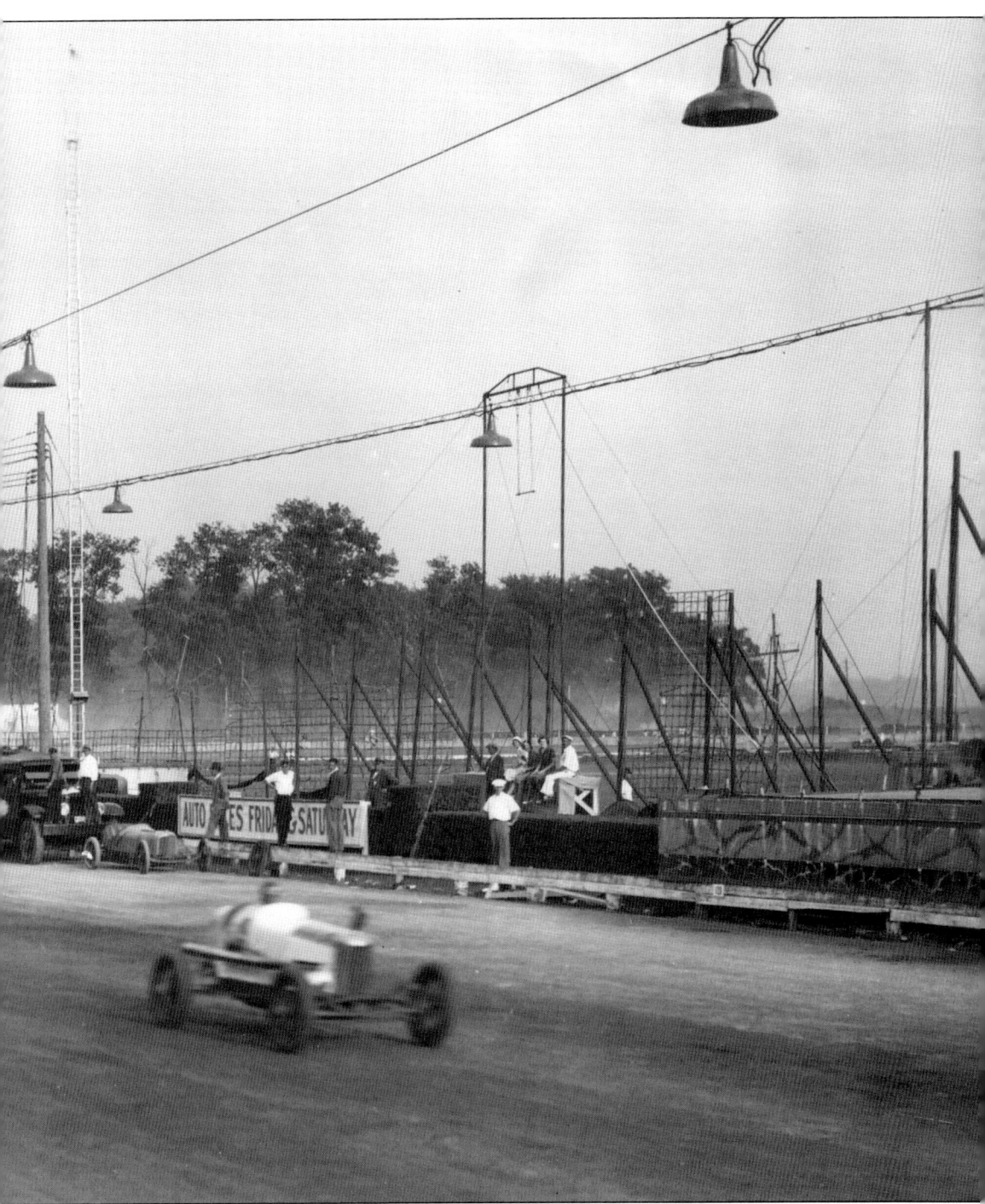

Gee Bee Model R1 flown by Jimmy Doolittle at the 1932 National Air Races in Cleveland, Ohio, which was on display outside the Industrial Arts Building. Earlier in the month, Doolittle had set a new world landplane speed record of 296 miles per hour in the craft at the air race's Shell Speed Dash.

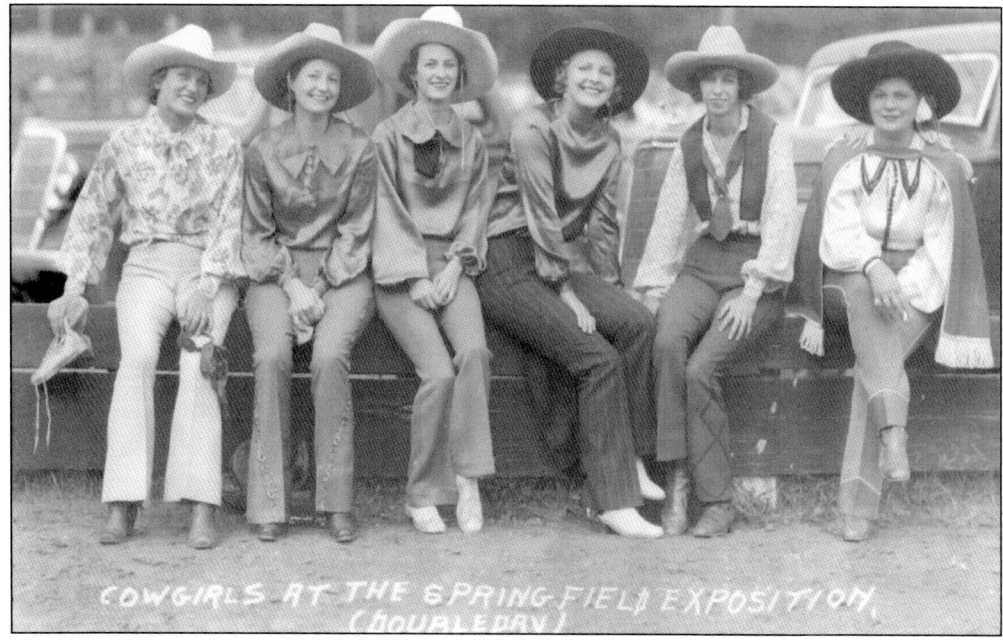

Billed by the exposition as the first and "most pretentious event of its type ever undertaken East of the Mississippi," the World's Championship Outdoor Stampede and Rodeo was performed twice daily from September 17 to 23, 1933. Two corrals were constructed on the racetrack infield, with six chutes leading to the racetrack arena. More than 100 cowgirls and cowboys took part in trick and fancy riding and roping, Roman, standing, and chuck wagon races, lasso contests, quadrilles on horseback, bronco riding, steer riding and bulldogging, and other "similar sports peculiar to the cowboy, the cattle range, and the round-up, which have made the stampedes of Calgary, Cheyenne, and Pendleton world famous." The above photograph features some of the show's cowgirls; the photograph below shows Hazel Moore riding Perfect Lady. Moore designed and made costumes for herself and others who rode with her. (Both, author's collection.)

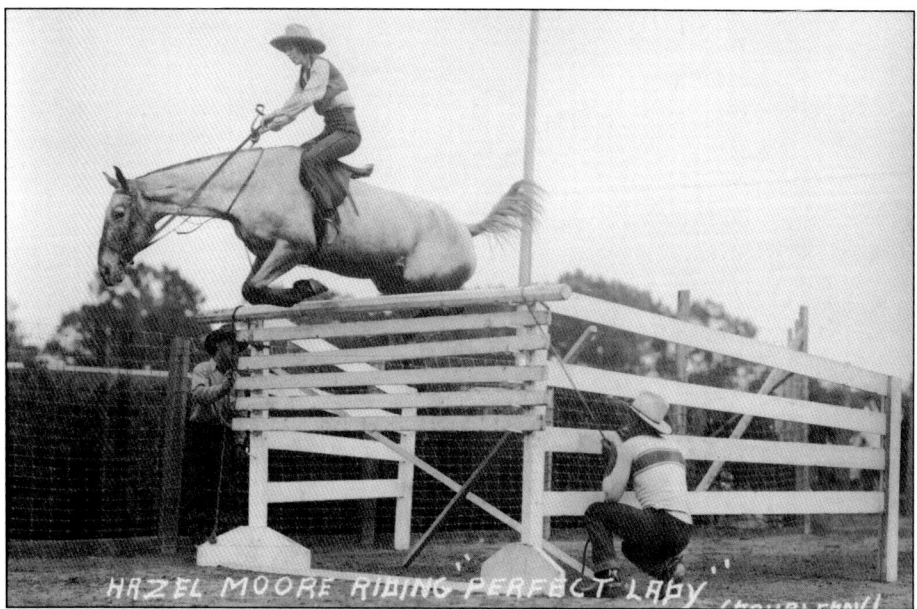

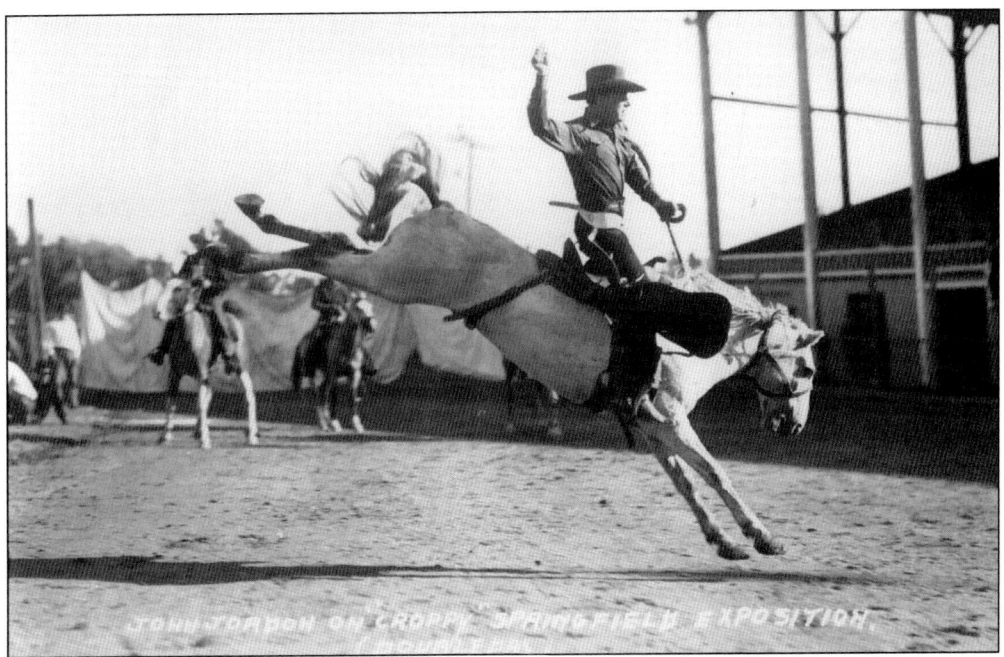

John Jordan started in rodeo work in 1922. He was the 1937 North American bronc riding champion and appeared in Western movies. He later owned and managed his own Wild West show. After suffering an injury in the arena, he became a rodeo announcer, including for many shows at Madison Square Garden in New York City. (Author's collection.)

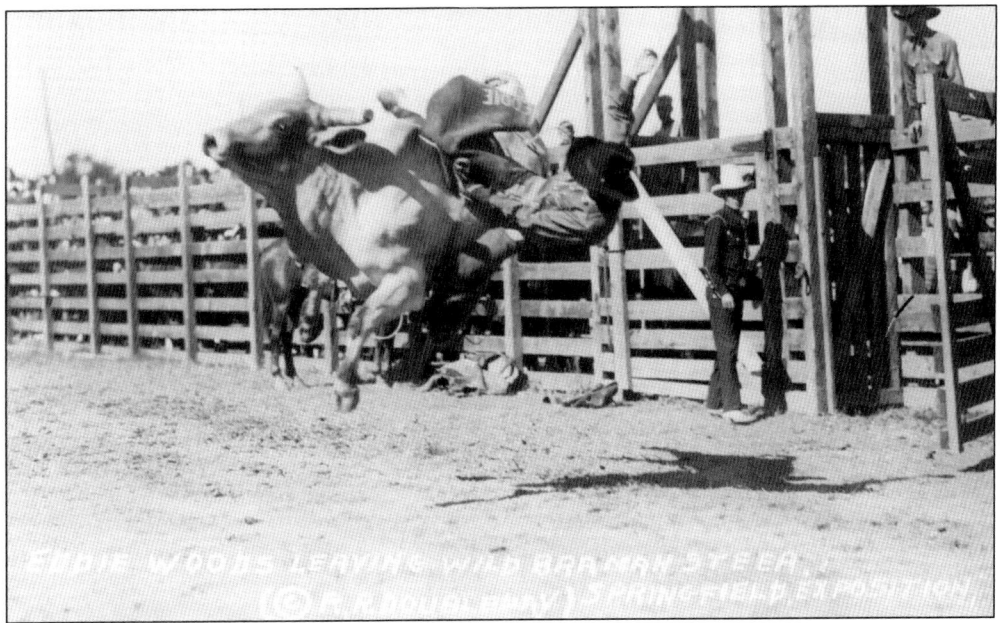

Eddie Woods finished first in the steer-riding contest at the 1933 exposition rodeo. The rodeo champion was featured in Camel cigarette advertisements and would go on to become bareback riding champion and bull-riding champion at the 1936 Reno Rodeo. That same year, he was elected vice president of the newly formed Cowboys Turtle Association, now known as the Professional Rodeo Cowboys Association. (Author's collection.)

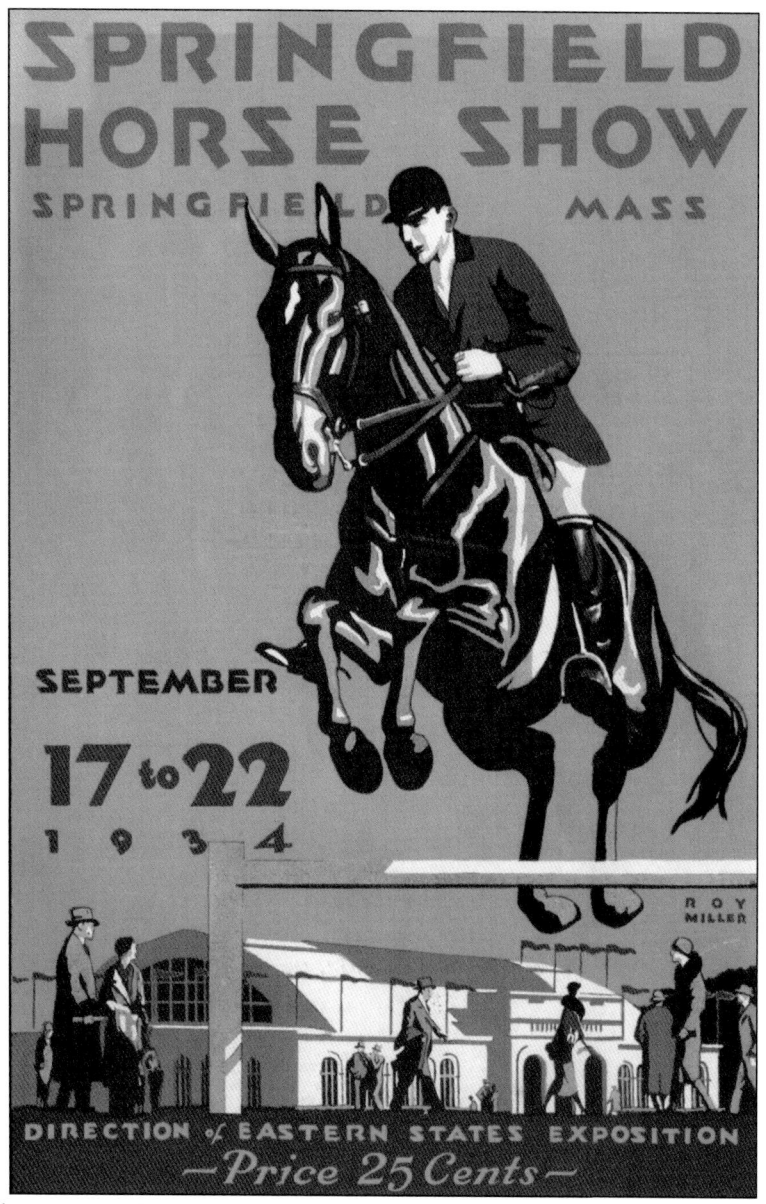

In spite of the worsening Depression, the 1933 Springfield Horse Show continued to be promoted as "the leading society event of the fall season." As was tradition, the horse show opened on Monday evening with Governors Night. The visiting governors of the 10 North Atlantic states, their parties, and several hundred other distinguished guests, along with the hosts assigned to them, representing local professional organizations and service clubs, were seated in specially decorated boxes in the Coliseum. Among that year's boxholders were Harry H. Caswell, general manager of W.F. Young, manufacturer of Absorbine Jr., and former president of the Advertising Club of Western Massachusetts; Mrs. M.J. Duryea, daughter-in-law of Charles Duryea; Massachusetts governor Joseph B. Ely; Harry G. Fisk of the Fisk Rubber Company; M. Robert Guggenheim, of the prominent Guggenheim family; aviatrix Maude Tait Moriarty; and Adrian Van Sinderen, a banker from New York City and president of the American Horse Shows Association for three decades. A breeder of ponies, his stables won more than 2,500 ribbons. (Author's collection.)

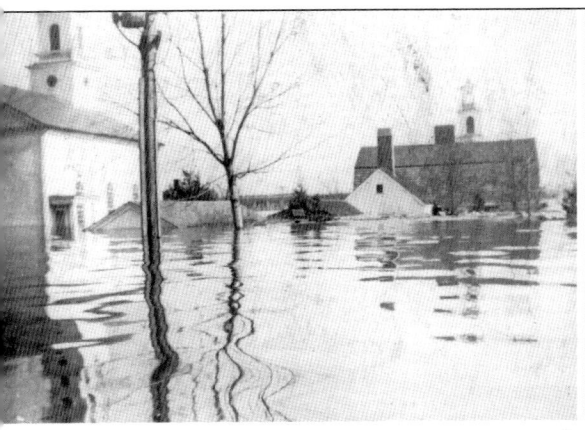
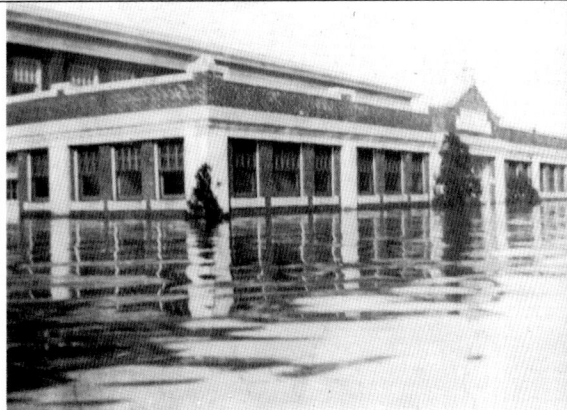
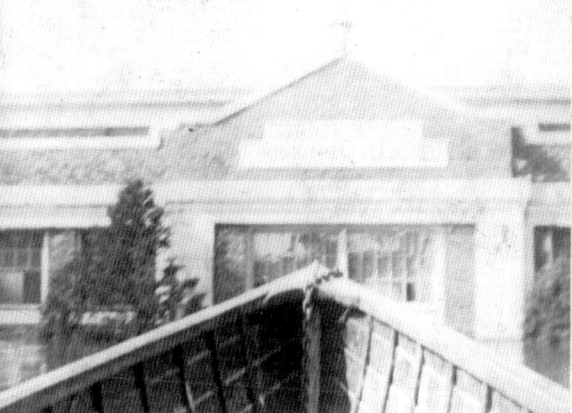
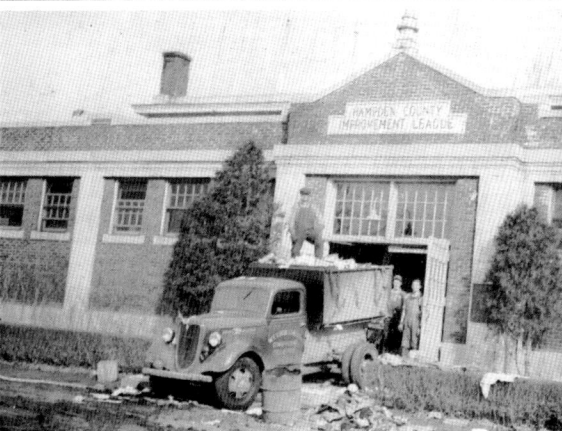

The March 1936 flood that inundated the Connecticut River Valley covered the exposition grounds with water up to 18 feet deep and caused $60,000 worth of damage (the equivalent of over $1 million in 2016). Exposition trustees voted to repair all damage immediately, and that fall's fair program proclaimed, "happily, the work of reconstruction has been finished, all traces of the flood have been removed, [and] many improvements have been made" in anticipation of the 20th anniversary fair that September. These images show, clockwise from top left, the view looking south across the grounds of the New England Colonial (Storrowton) Village, the Hampden County Improvement League building reflected in floodwaters that reached its windowsills, dump trucks hauling away the tons of mud and water-damaged material after the waters receded, and the view of the League Building at the height of the flood taken from a rowboat by Robert "Bob" Allen, who was custodian of the building from 1933 to 1967. (All courtesy of Linda Noftall.)

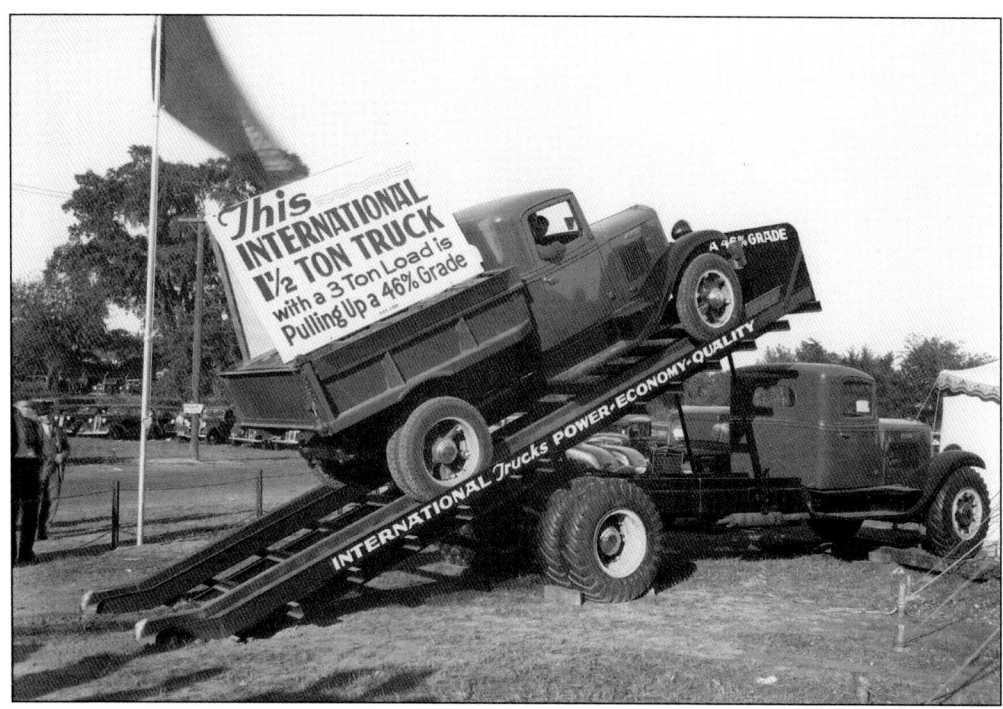

Farmers in general, and probably New England farmers in particular, are mostly a stubborn lot and not easily convinced—of anything. What better way to show the merits of a new truck than actually putting it into use? Of course, if one did not have a need to pull a three-ton load up a 46-percent grade, it might have taken a little more effort to make the sale.

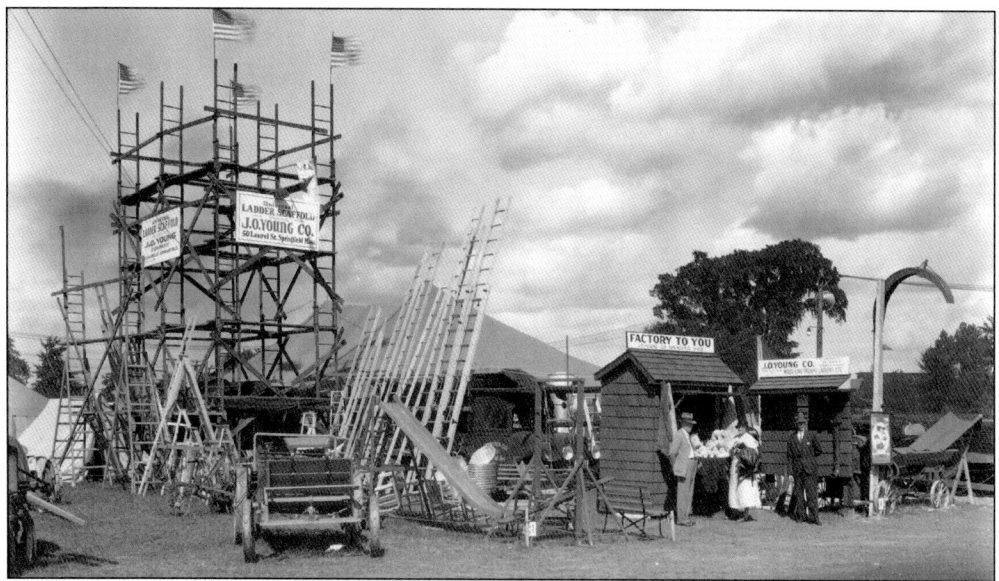

The J.O. Young Company exhibited continuously at the exposition from 1932 through 1941. Founded in 1900 by Judson Obediah Young and originally located in Springfield as a manufacturer of scaffolding, ladders, and eave troughs, the company later transitioned to installation of roofing and siding. In 1970, the business moved to Agawam, where it remains to this day as general contractors for construction, remodeling, and home improvement projects.

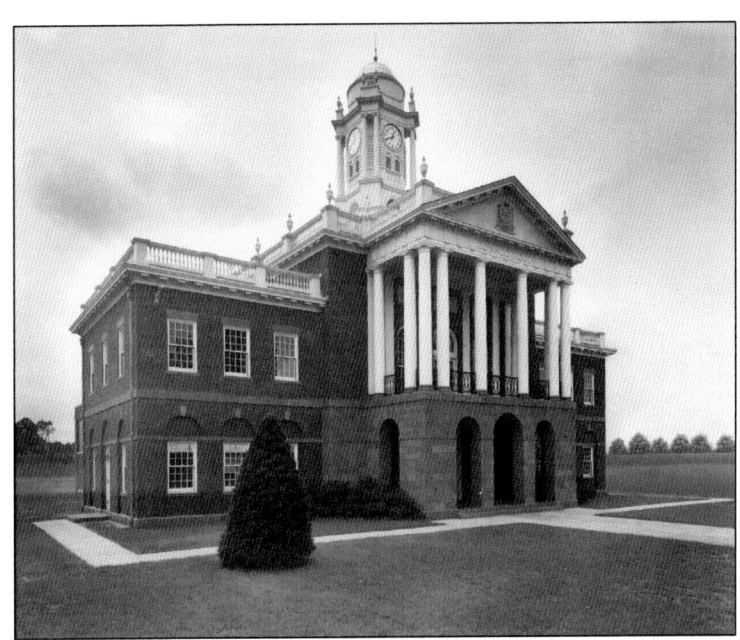

The Connecticut State Building was dedicated on September 19, 1939, with former Connecticut governor Wilbur L. Cross and then governor Raymond E. Baldwin the featured speakers. A crowd of 28,886 people, though larger than that of the same day the previous year, was described in the *Springfield Republican* as looking "very small—as it always does on Tuesday of Exposition Week—following the mobs of Children's Day."

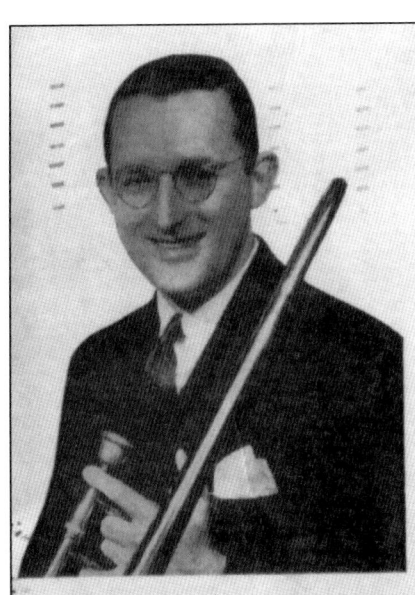

"Radical" changes were made to the 1939 musical program with the addition of a two-and-a-half-hour stage show featuring Tommy Dorsey and his orchestra. The exposition also added "as an extra attraction a championship jitterbug contest for the 'Tommy Dorsey Cup' in which the youngsters will really show what happens when a top flight swing band goes into action." (Author's collection.)

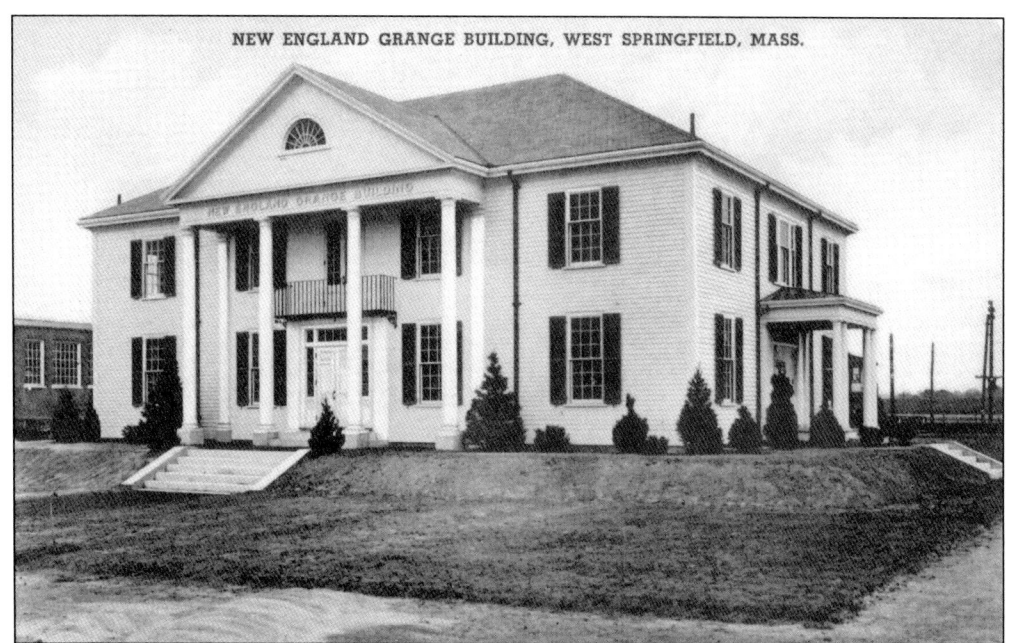

The dedication of the New England Grange Building took place on Wednesday, September 21, 1938, a day most remember for another reason. The building was designed by architect Joseph Chandler of Boston and built by Grange member Willard T. Kelly of Merrimac, Massachusetts. The building contained 16,000 bricks, 32,832 feet of framing lumber, 2,600 pounds of nails, and 9,654 square feet of Celotex. (Author's collection.)

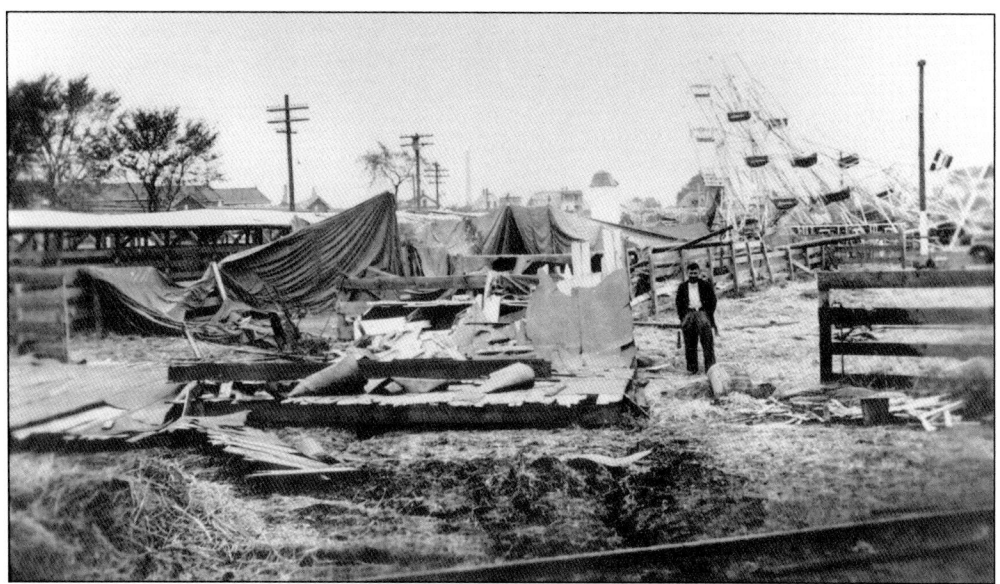

Rain fell steadily as the 1938 fair opened, turning the grounds to mud and causing outdoor exhibits to close. Flood controls installed since 1927 prevented another inundation, but the winds of the 1938 hurricane, which arrived on September 21, wreaked havoc on the exposition grounds, uprooting trees, smashing windows, tearing roofs off buildings, and famously turning the Ferris wheel into a mass of twisted metal. Miraculously, there were no serious injuries.

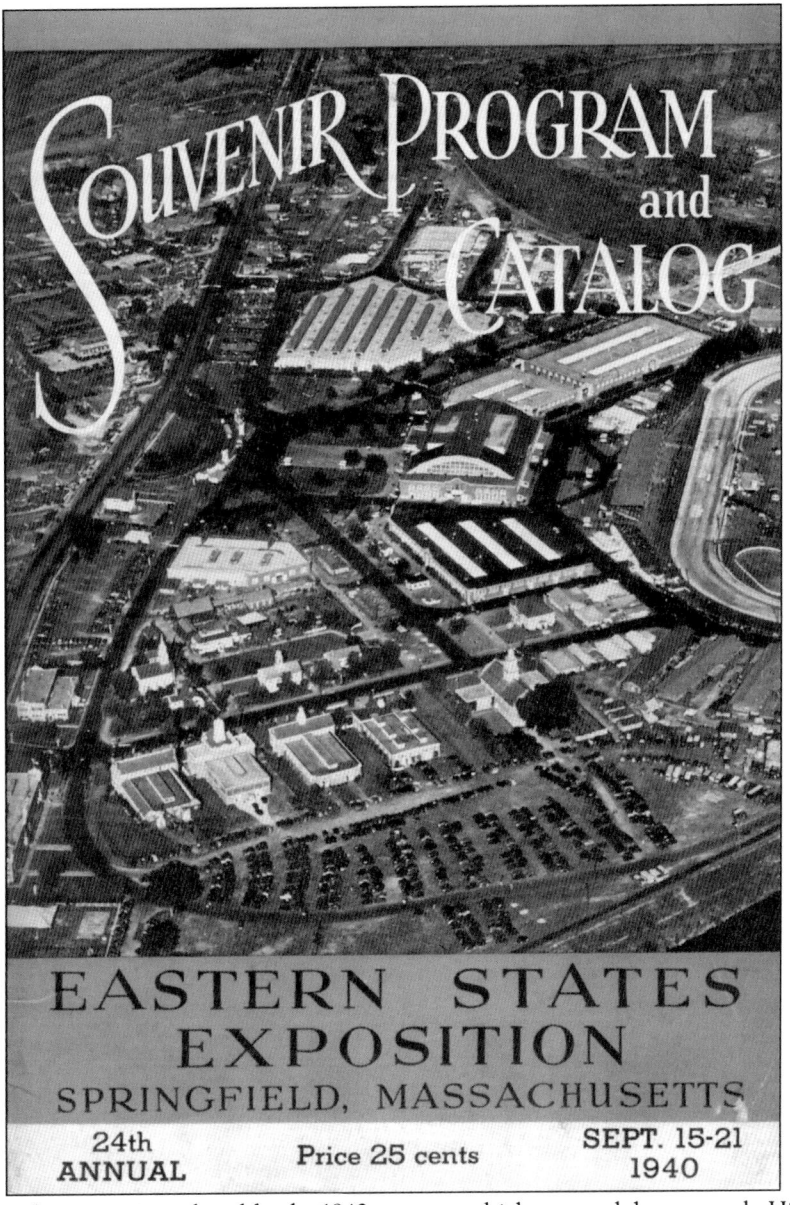

A new cover design was introduced for the 1940 program, which was used that year only. Highlighted was an updated aerial view that showed the previous decade's additions to the grounds: three new state buildings—Vermont (1929), New Hampshire (1930), and Connecticut (1939)—as well as Storrowton Village (1930) and the New England Grange Building (1938). New for 1940 was the introduction of motorcycle racing; motorcycles had raced on the grounds before—most notably in 1927 when the three- and 10-mile national championships were held there—but never during Exposition Week. Bill Huber won the 10-mile event on his Harley Davidson; the majority of those participating rode Indians, the local favorite. Also introduced that year were the Exposition Follies of 1940. This all-star musical extravaganza involved a cast of 150 on an expansive 140-feet-wide by 90-feet-deep outdoor stage. Accompanied by a 21-piece orchestra, the Exposition Follies featured 24 dancing girls, singing, high-wire acts, acrobatics, animal comedy acts, a cowboy quartet, dramatic lighting, and a dazzling fireworks finale each evening. (Author's collection.)

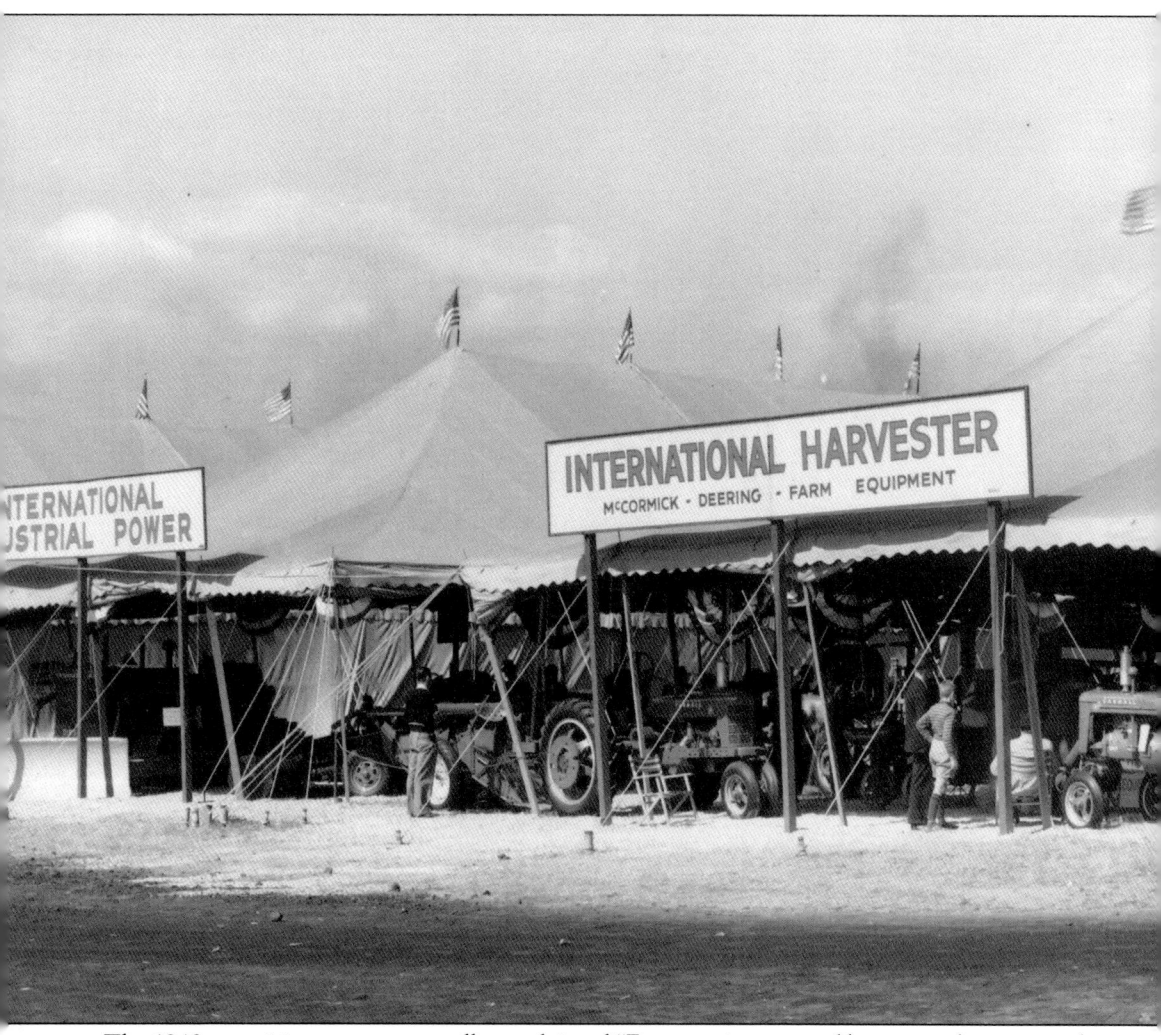

The 1940 exposition program proudly proclaimed "Farm equipment and heavy machinery worth $1,750,000 (the equivalent of $29.5 million in 2016) will be on display in the largest farm machinery and industrial show ever assembled in New England. In addition to the huge Industrial Arts Building it has been necessary to set aside seven acres of outdoors exhibition space." Much of the outdoor exhibits were housed under huge tents, such as these containing the International

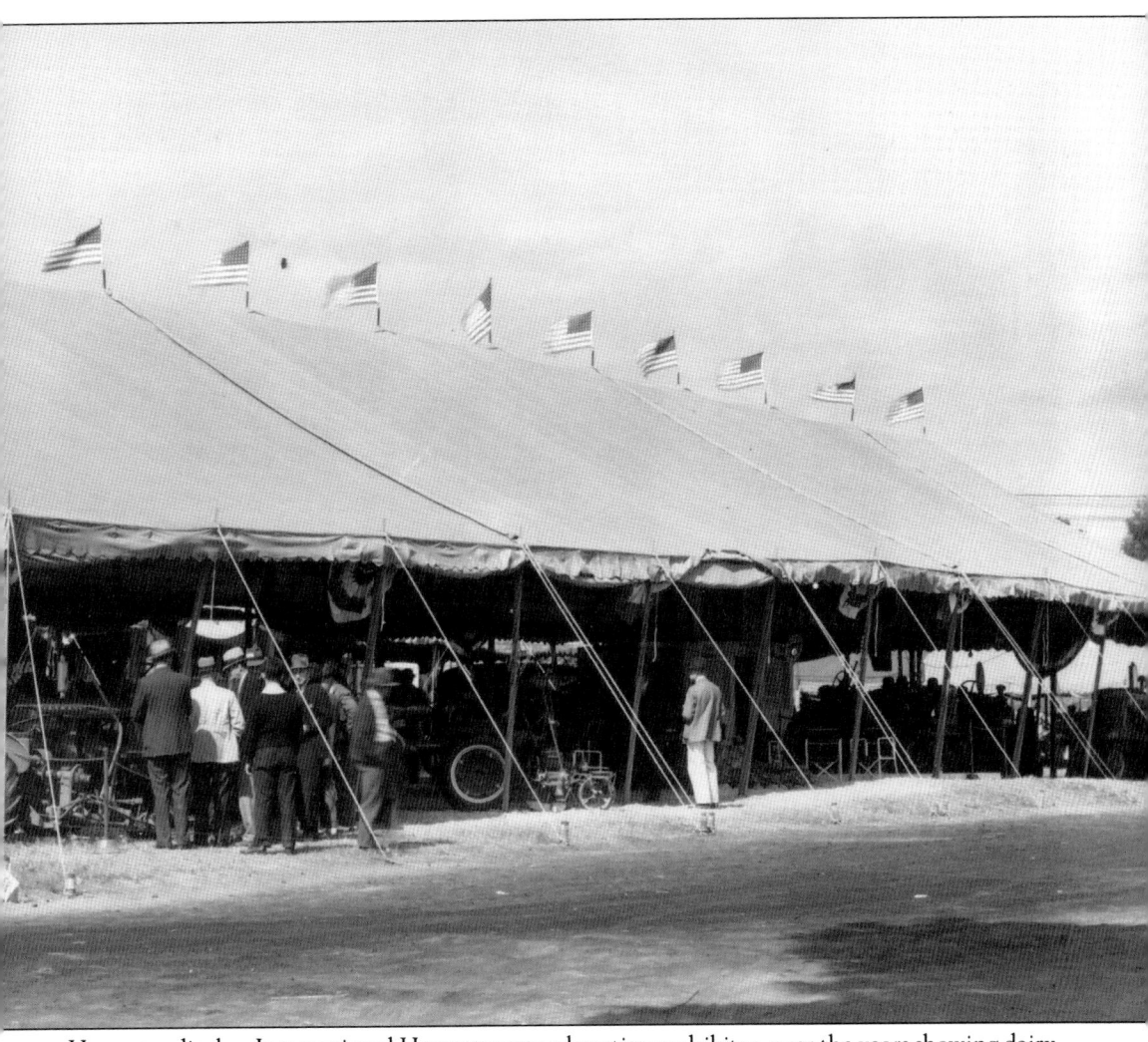

Harvester display. International Harvester was a longtime exhibitor, over the years showing dairy equipment, trucks, tractors, implements, and household refrigerators and freezers. Seen here under the tent are the all-purpose Farmall H, center left, in production from 1939 to 1953, and at center right, the smaller Farmall A, manufactured from 1939 to 1947. The model seen here features the offset Cultivision engine arrangement.

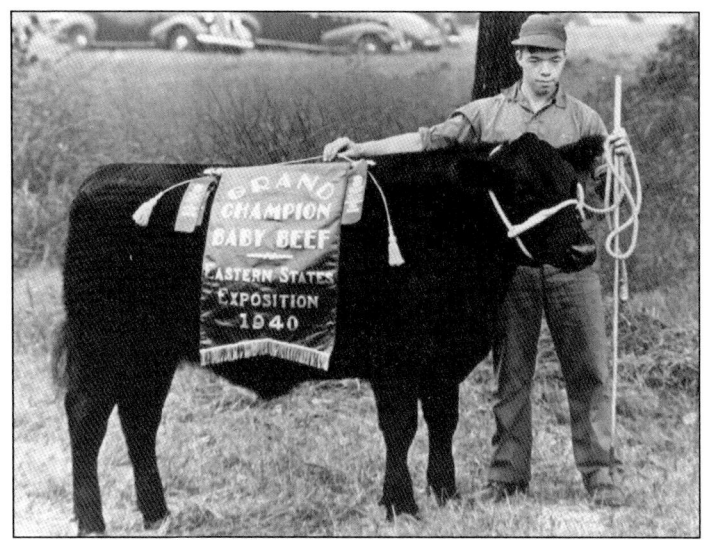

Lee Jenks of Feeding Hills, Massachusetts, raised the 1940 grand champion Baby Beef, receiving $1,030 at auction for the steer. Baby Beef exhibitors are immensely proud of their achievements—Jenks's 2013 obituary listed few things: his survivors, his World War II prisoner of war status, his receipt of the Purple Heart and French Legion of Honor, and his 1940 grand champion Baby Beef. (Courtesy of Bill Fearn.)

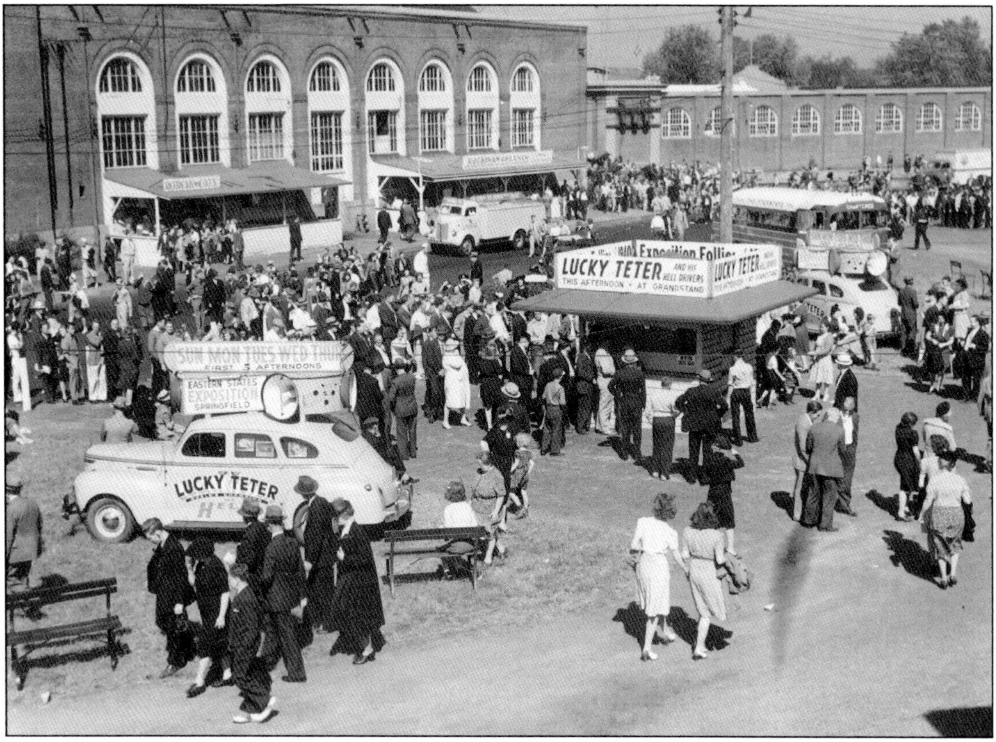

Automobile thrill shows were a regular feature at the exposition for more than four decades, beginning with Lucky Teter and his Hell Drivers in 1936. Other stunt drivers who appeared at the track over the years include Jimmy Lynch and his Death Dodgers, Joie Chitwood's Auto Dare Devils and later his Chevy Thunder Show, Jack Kochman's Hell Drivers, Dan Fleenor and his Hurricane Hell Drivers, and Charlie Belknap's Hollywood Stunt Show.

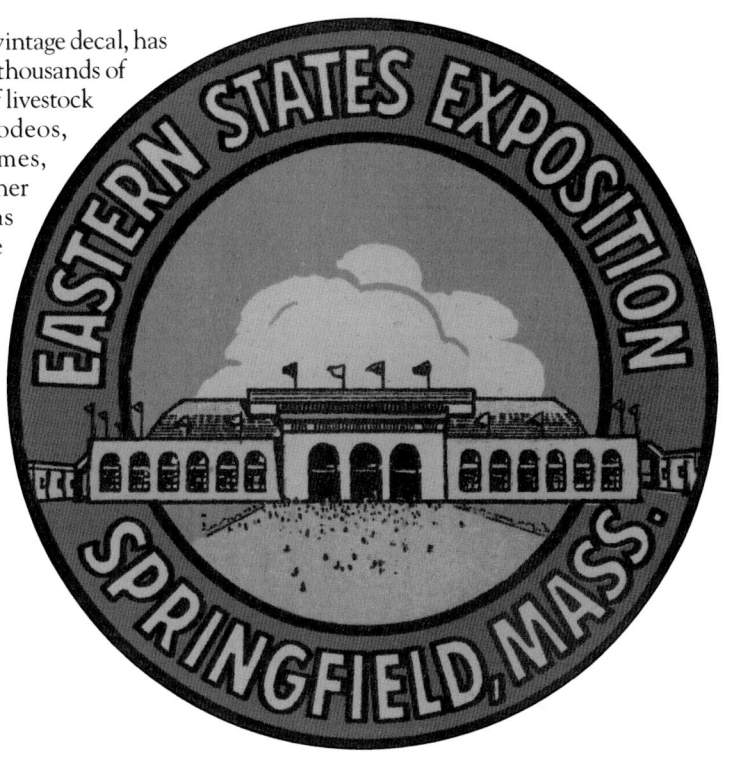

The "Colly," seen here on a vintage decal, has played a part in the lives of thousands of local residents. The scene of livestock judging, horse shows, rodeos, graduations, hockey games, circuses, concerts, and other events, the Coliseum was also the fitting setting of the exposition's Community Breakfasts leading up to the fair's centennial. (Author's collection.)

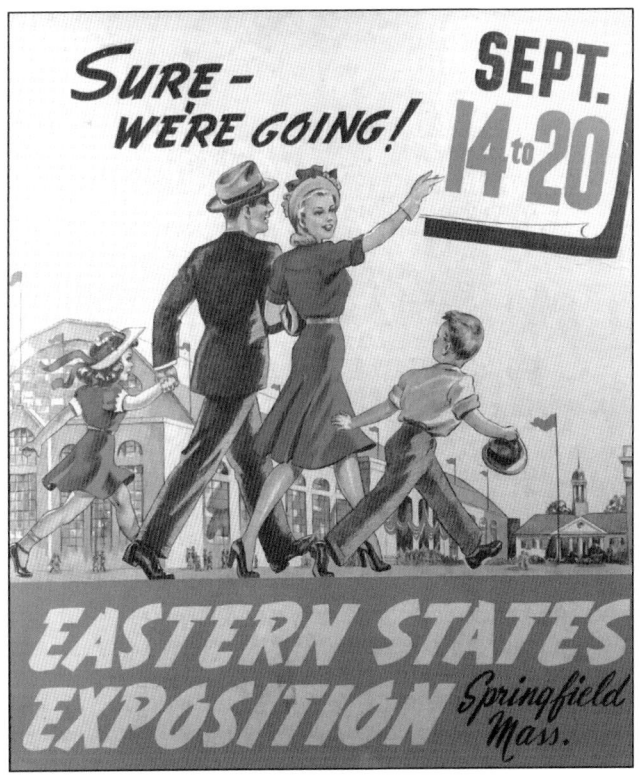

"Sure–we're going!" The clouds of war had not yet quite reached Main Street, USA; Pearl Harbor was still months away. Poster artwork for 1941 featured a fashionable family striding confidently across the exposition grounds, with the Coliseum prominently visible in the background. At the end of Commonwealth Avenue there stands what looks like the Maine Building where the Massachusetts Building should appear. Perhaps the illustrator was a "Mainiac." (Author's collection.)

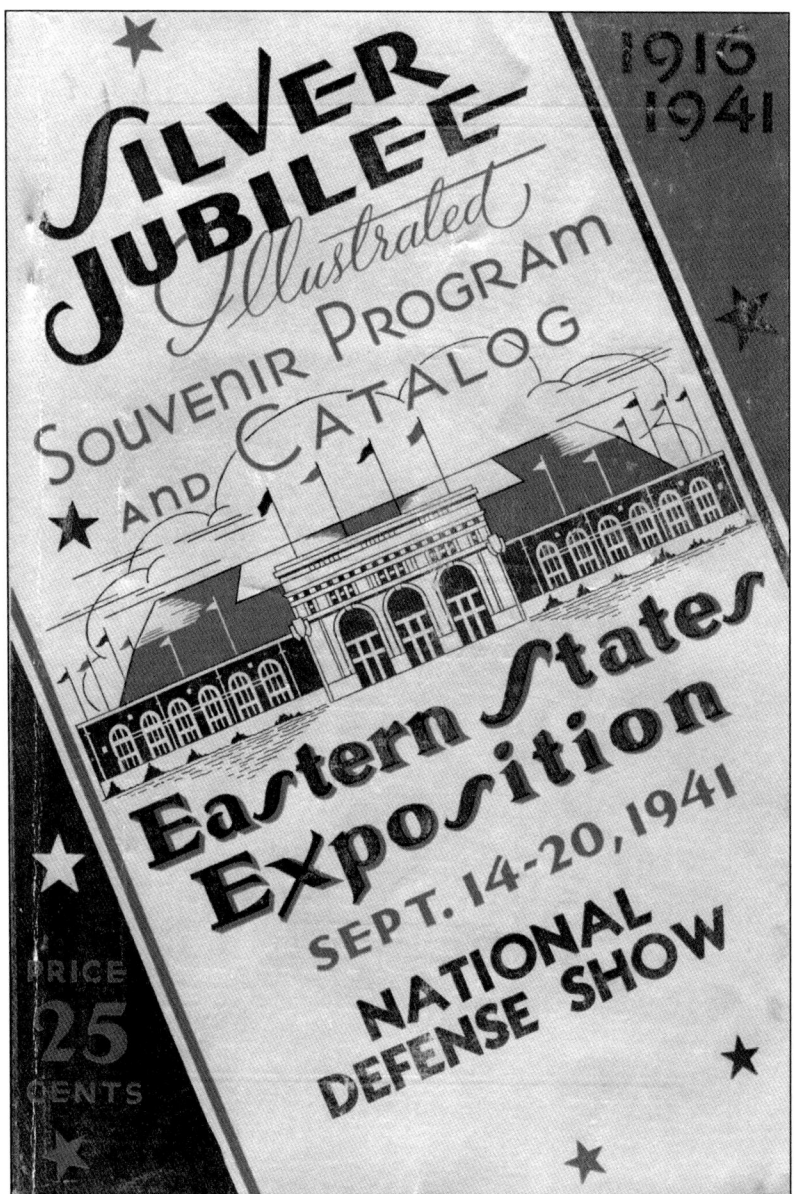

With war raging in Europe and the US government "preparing to defend this hemisphere," as described in the 1941 program, the 25th anniversary Silver Jubilee of the Eastern States Exposition shared the spotlight with 1941's National Defense Exposition and Industrial Show. To prepare for the show, the 68th Coast Artillery Regiment, under orders of the War Department, transported their materiel to the exposition from Fort Devens in a 14-mile-long, 388-vehicle convoy. The regiment's 1,839 men and 82 officers camped on the exposition grounds for the duration of the week, transforming the fair into an armed camp. Featured every evening at 10:30 p.m. was the "air defense spectacle." Batteries of antiaircraft guns were moved to the infield of the racetrack, and as air raid sirens wailed, the track area was blacked out and the regiment's 15 giant searchlights turned on to illuminate the skies above West Springfield, their 1.2 billion candlepower beams clearly visible for more than 50 miles. Armed forces recruiters were also present and prepared to enlist men at the exposition. (Author's collection.)

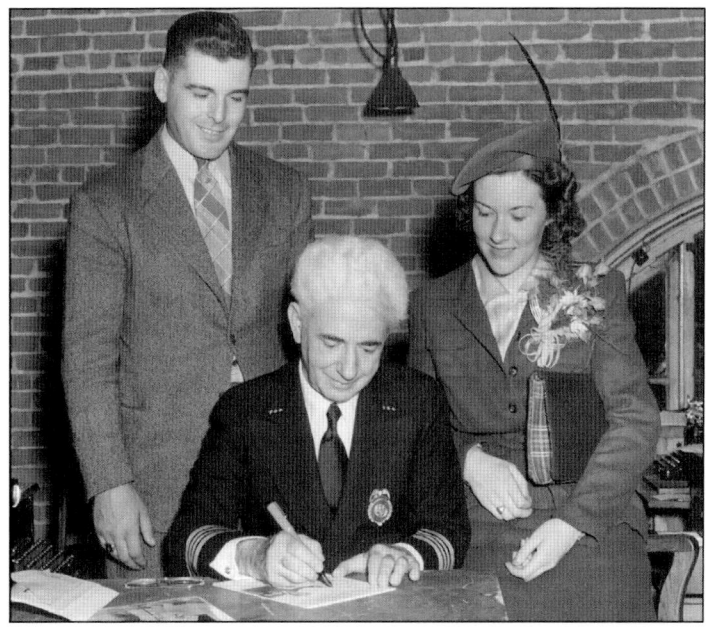

Selected by the Maine Development Commission to represent the Pine State as the ideal farm couple, George S. and Kathleen Bishop of Fort Fairfield, Maine, receive bandleader Edwin Franko Goldwin's autograph at the 1941 fair. The Bishops were guests of the exposition, in recognition of the importance of agriculture in New England. They took part in the governor's reception parade in the Coliseum; Kathleen was also named the fair's Silver Jubilee girl.

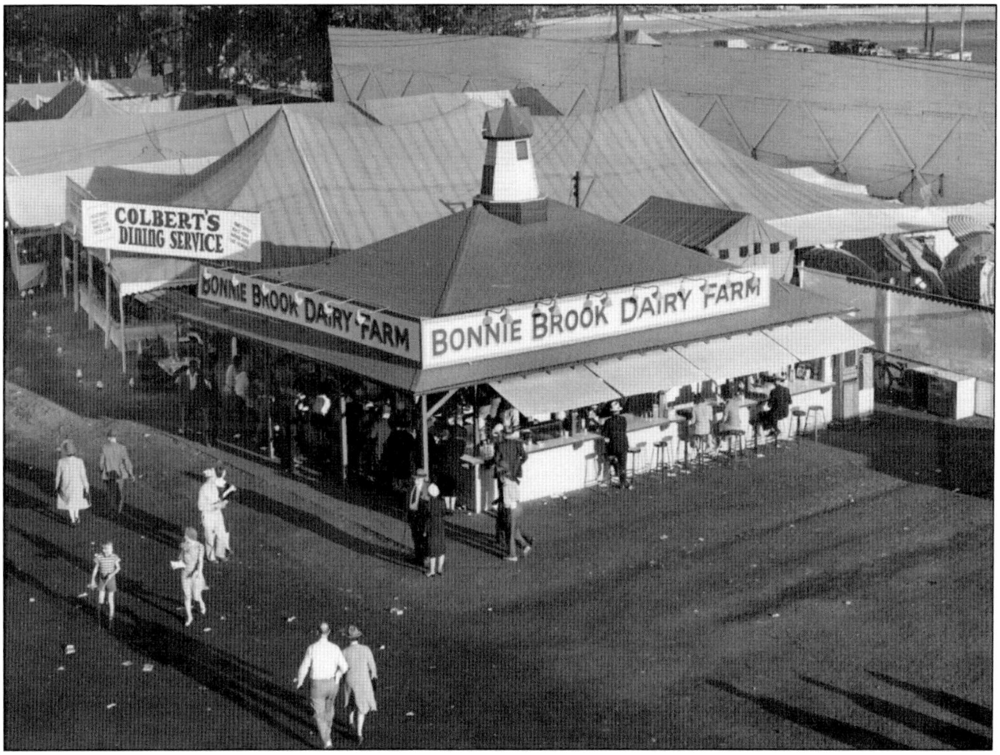

Several dairy farms with the name Bonnie Brook were located within the eastern states, in Bloomfield, Connecticut; Pawtucket, Rhode Island; and South Sudbury, Massachusetts. While no definitive information could be found, it is likely that this dairy bar was run by H.P. Hood, who had purchased the Sudbury dairy in the 1920s and operated it as a model farm. Bonnie Brook Dairy was not listed as an exposition concessionaire, but H.P. Hood was.

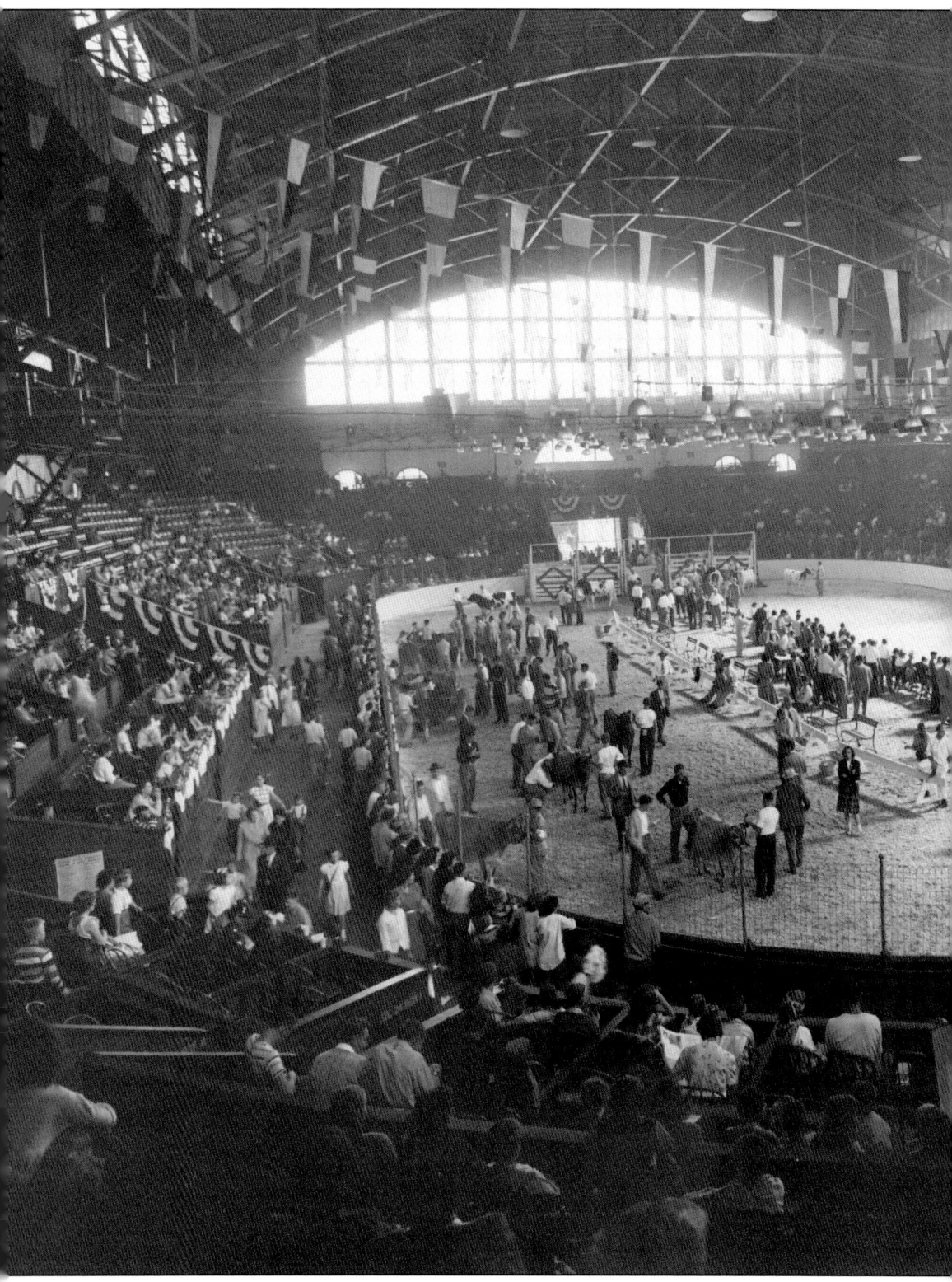

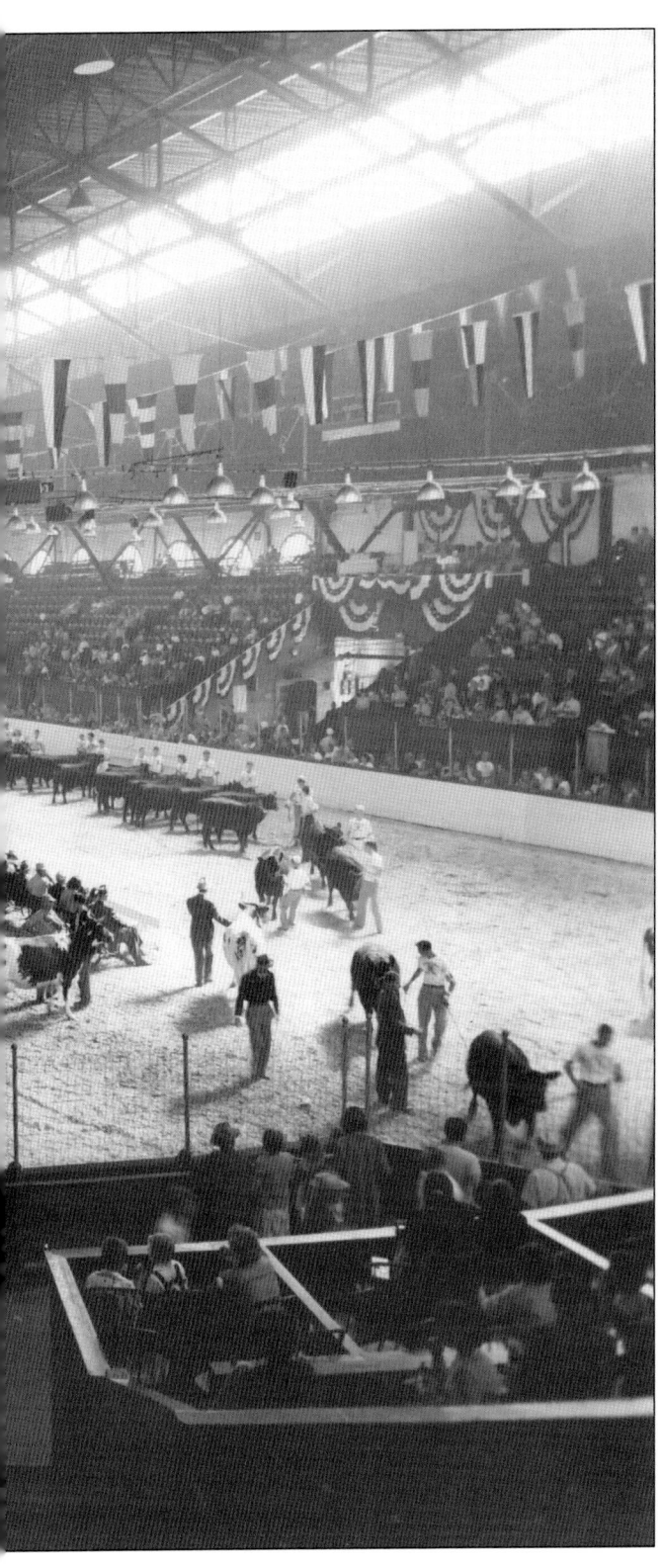

In 1941, the exposition called the Coliseum "one of the finest and most used structures on the exposition grounds. Of brick and steel construction, it has a high arched roof of glass that permits adequate natural lighting in the daytime, while powerful floodlights give it brilliant illumination at night. The building is 300 feet long and 200 feet wide, with a center show arena that measures 200 by 100 feet. It seats 5,600 persons. A wide concourse between the seating area and the arena fence allows free movement of foot traffic and provides space for thousands of spectators to stand. During the week of the Eastern States Exposition the Coliseum is used through the daytime for cattle judging." Seen here before the windows were covered, this photograph clearly illustrates that description.

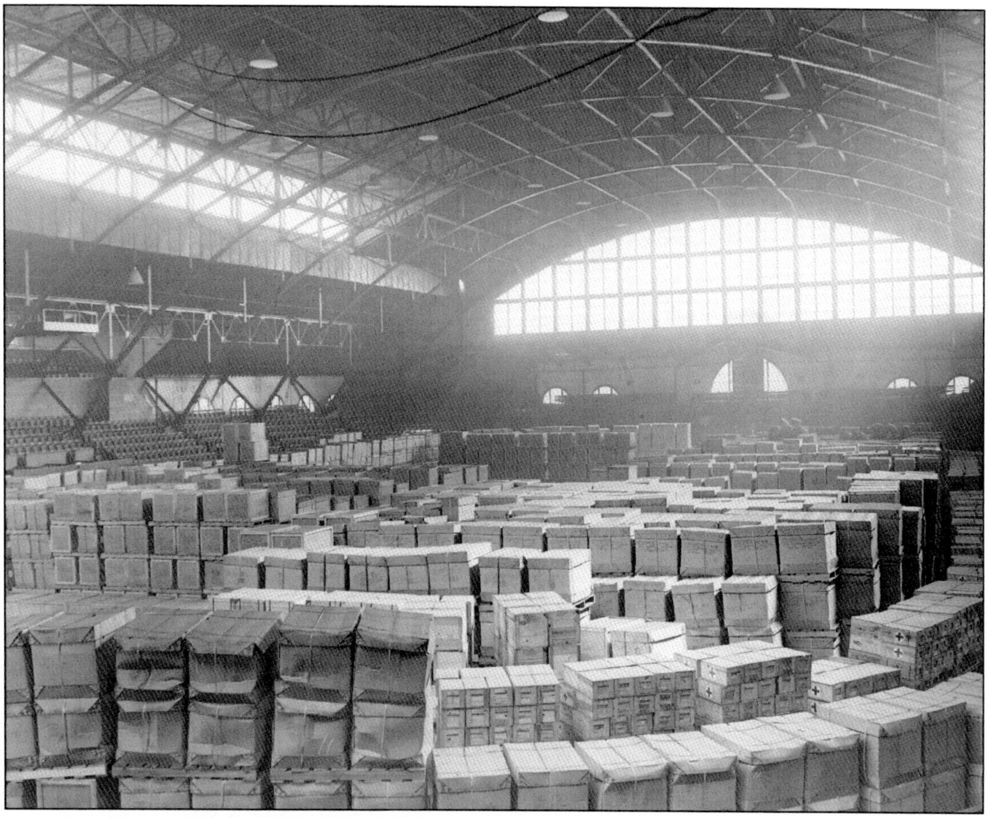

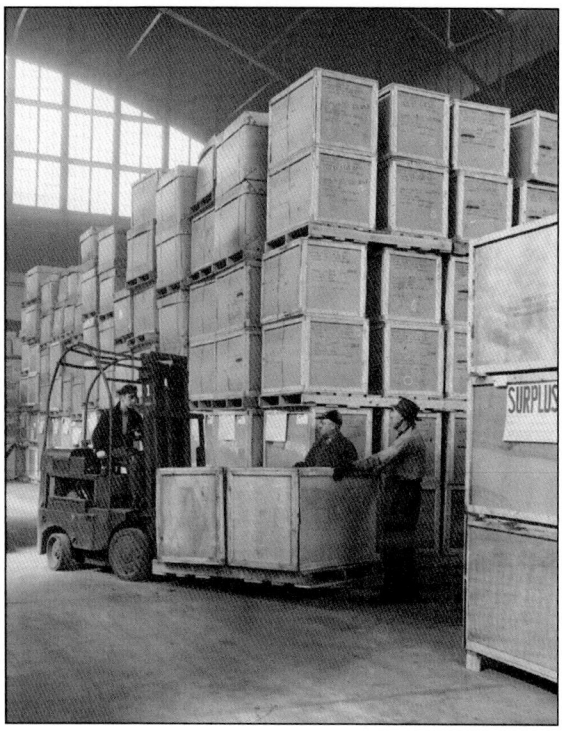

War was declared just months after the close of the 1941 exposition, and by September 1942, the entire exposition property was under control of the US Army, Philadelphia Quartermaster Depot, as part of a nationwide storage system for military textiles. There would be no fair that year or the next four, with millions of yards of wool fabric on site, military police in the Junior Achievement building, and Army officers quartered at Storrowton Village. At war's end, the exposition remained under government control as a surplus depot until early 1947. Despite the need to repair extensive damage caused to the grounds during its occupation, the 26th edition of the exposition opened as scheduled on September 14, 1947. Fears of waning public interest due to the five preceding exposition-free years were put to rest when a record 83,402 people arrived on opening day.

Three

SHOW WINDOW OF THE EAST

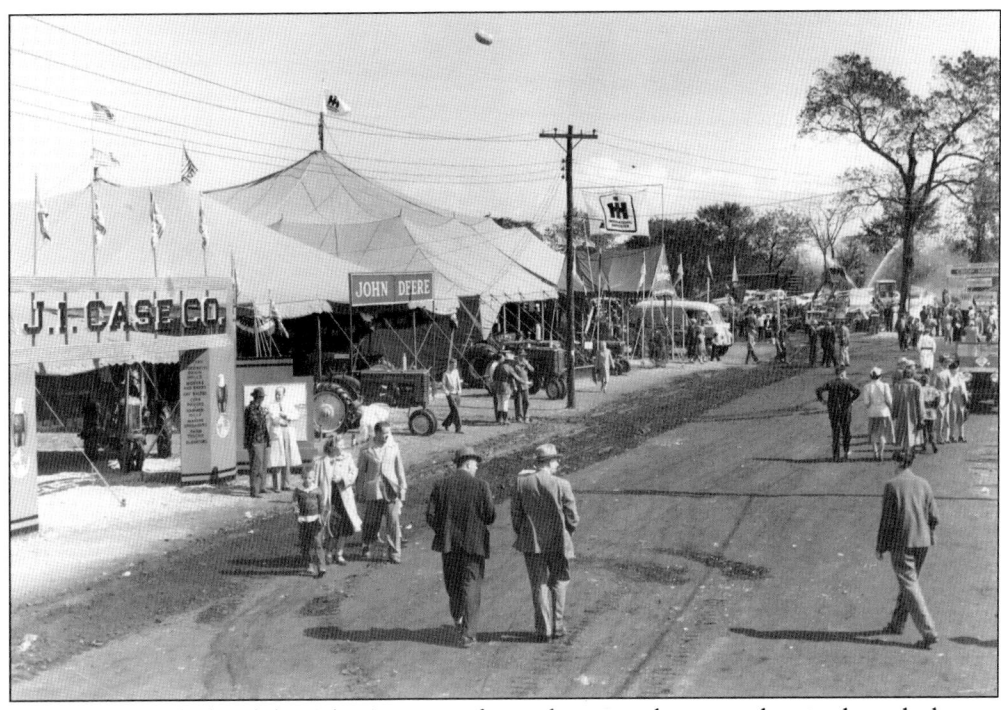

The exposition declared that "the American farmer has proved again and again through the war years that he can do *anything* in the way of crop and livestock production because he has *everything* to work with, either on his own farm or almost instantly available" as it announced "the biggest display of farm equipment and machinery ever seen in New England" for the 1947 show.

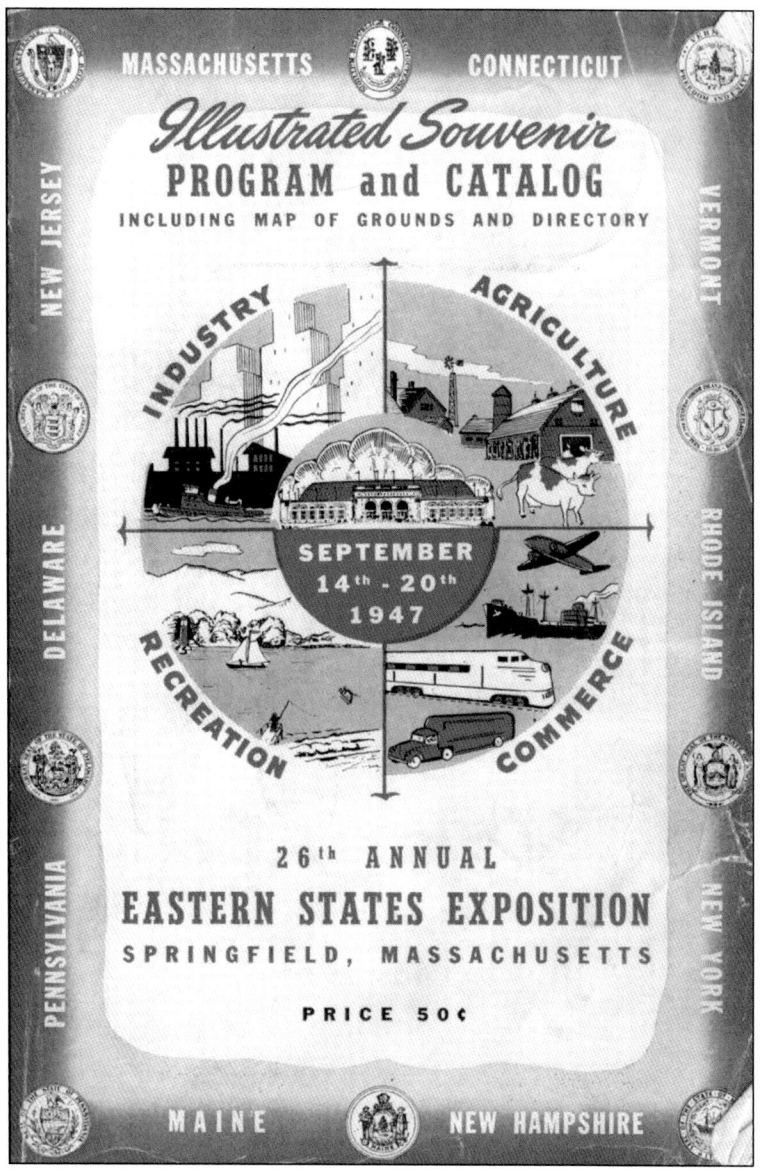

In the first postwar exposition program, notice was made of the revolutionary changes that had come about, "world-shaking in nature and which must completely alter the thoughts and acts of nations, continents, and their inhabitants toward one another. We have seen destruction and waste of natural resources and life which transcend any period in world history. Yet from this chaotic maelstrom there emerges one fact clearly and undeniably—life itself depends upon natural resources and agriculture. Nowhere else in the world has agriculture been developed to such heights of efficiency in production as in America. Nowhere else in the world do the tillers of the soil and breeders of livestock have so much machinery and equipment, so much scientific research and assistance at their disposal." Fair officials reaffirmed that the exposition, "dedicated since its inception to the advancement of agricultural and industrial interests of New England and adjacent states, is again prepared to resume its leadership in bringing together in one place the greatest array of exhibits and demonstrations of agricultural, industrial, recreational and cultural displays in its history." (Author's collection.)

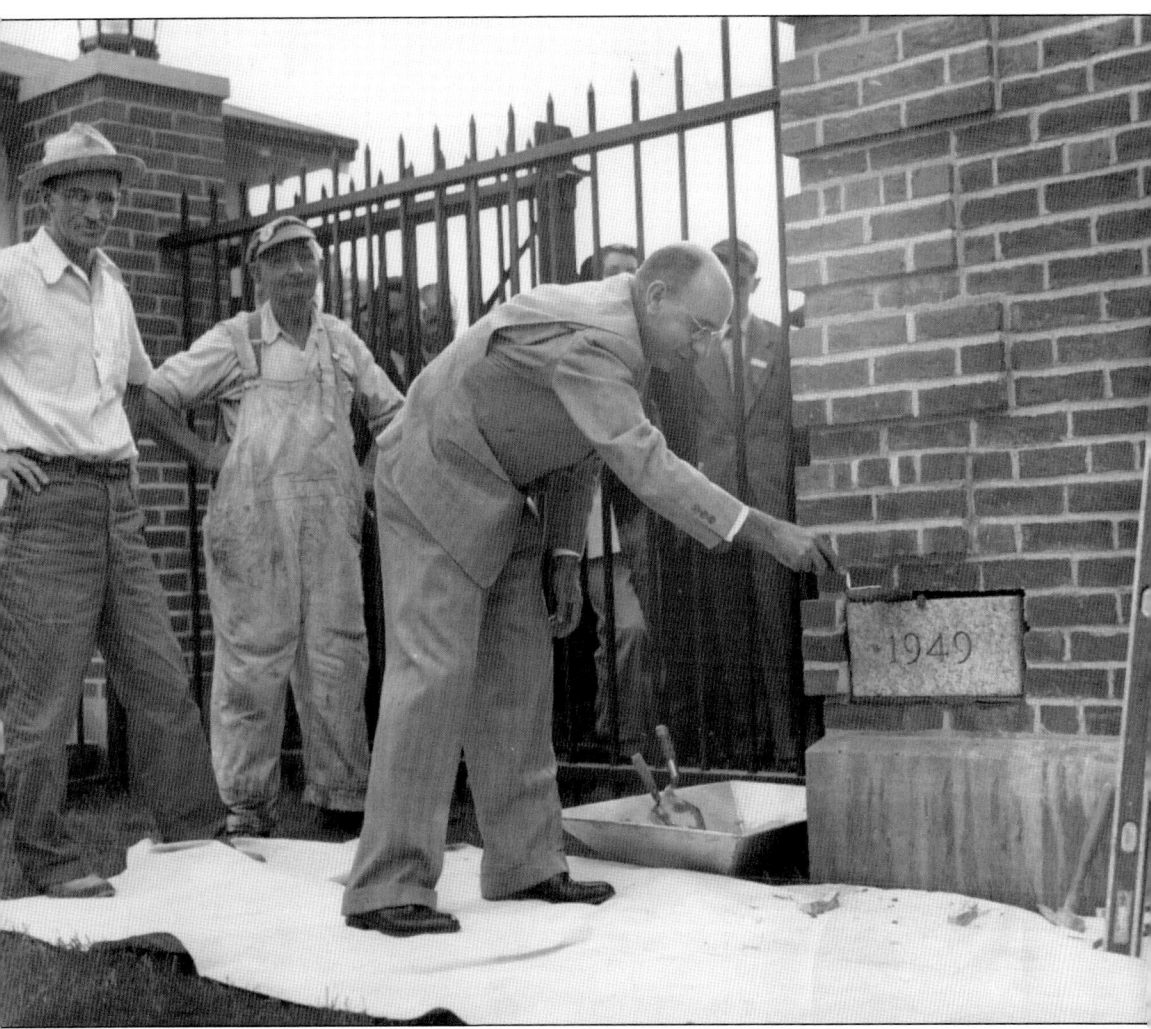

Prior to 1949, the administrative offices of the exposition were located in downtown Springfield except during Exposition Week, when they occupied temporary quarters in the Coliseum. Founder Joshua Brooks felt that in its early years, the "city" address would lend credibility to the organization, a point well-considered, given the progressive and prominent reputation of Springfield at that time. However, after nearly 30 years of operation and wide influence as an important and well-established event, the exposition sought to consolidate operations in West Springfield. The new building, designed to complement structures previously constructed on the grounds, was dedicated on September 19, 1949, as part of that year's Governors Day program. Named in memory of exposition founder Joshua L. Brooks, who had passed away earlier that year, on February 16, the cornerstone was laid by his son J. Loring Brooks Jr., who later served as exposition president from 1958 to 1967. The Brooks Building continues to serve as the organization's headquarters.

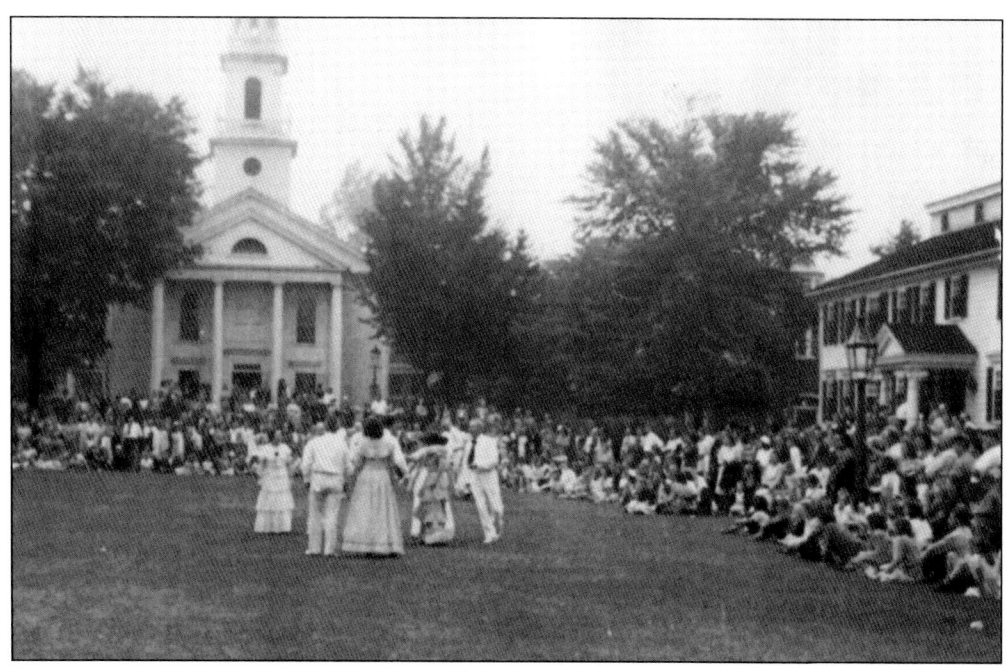

Square dancing has had a presence at the exposition since 1930, when Katherine Heron, director of the Home Department, instituted the "old time" dances in the ballroom of the Atkinson Tavern at the New England Colonial (now Storrowton) Village. Over the years, Storrowton dancers have performed at Radio City Music Hall; Boston's Symphony Hall; in Washington, DC, for three first ladies; and appeared in *National Geographic* magazine. (Author's collection.)

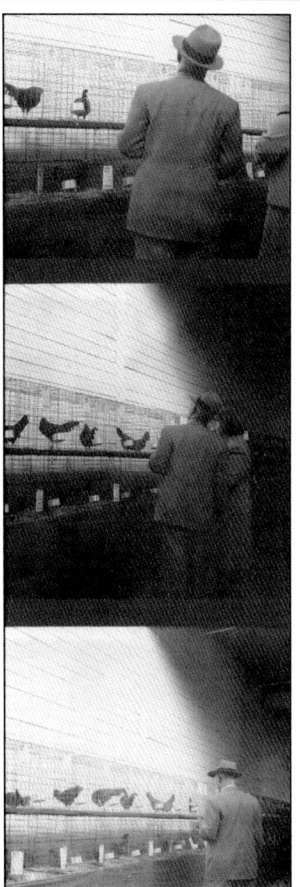

The reputation of the exposition's poultry show as a sizable and comprehensive event with high-caliber judges attracted more than 2,000 entries from breeders in New England and across the country in 1950. Local winners in the show included Charles J. Johnson and Alvin Rich of West Springfield, Edward Wollner of Feeding Hills, Ernest G. Freeman and Kenneth Hinshaw of Agawam, and Dorothy E. Dickinson of Hampden. (Author's collection.)

Farm machinery production reached record levels in 1948, and New England farmers enjoyed their most prosperous year yet, with farm product sales of over $789 million. The following year, demand for outdoor display space also increased, resulting in the largest farm machinery show in the east. West Springfield farmer Leonard Lapinski recalls that east of the large display area was a demonstration field where farm machinery was put into use.

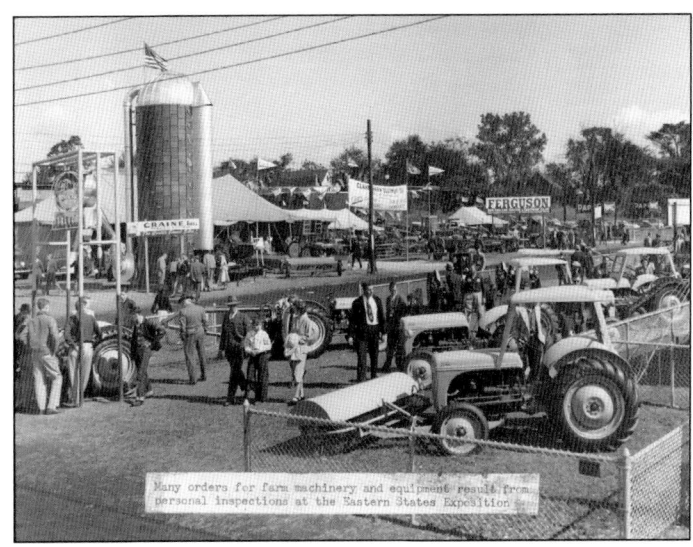

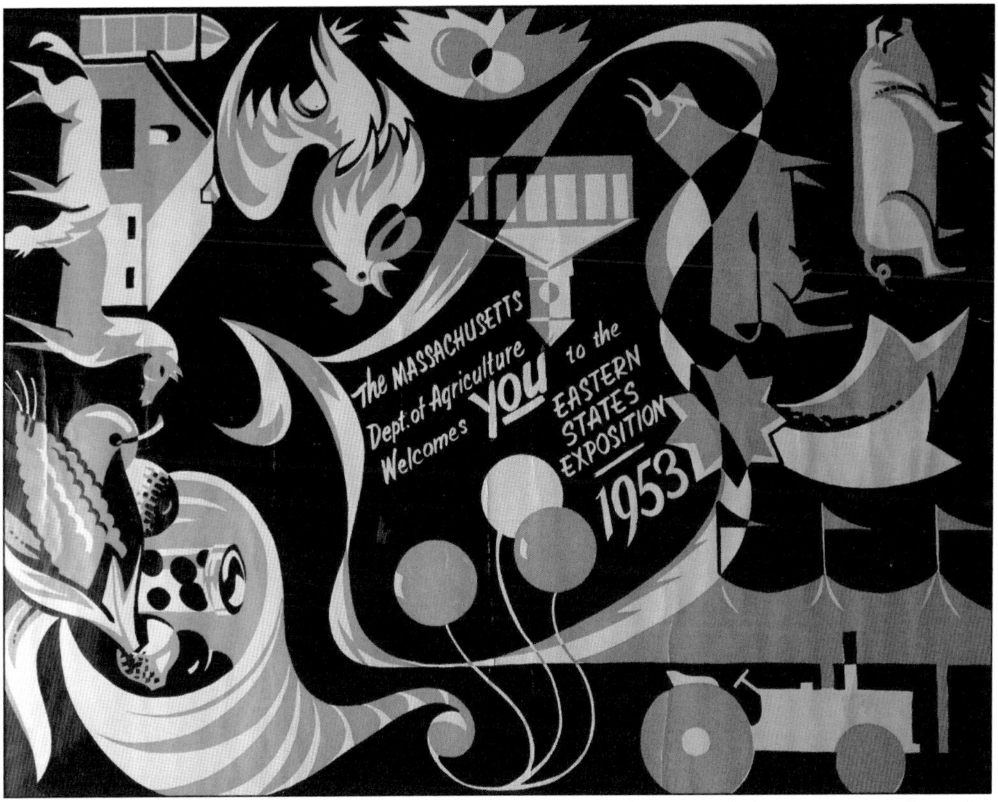

This poster from 1953 visually references many important features of the exposition, from the $5 million farm machinery display, to the poultry, cattle, swine and horse shows, to native produce and Storrowton Village. The year 1953 was the second year of an eight-day schedule that included two full Sundays; prior to 1952, the seven-day fair opened on a Sunday afternoon (Music Day) and ended on Saturday evening. (Author's collection.)

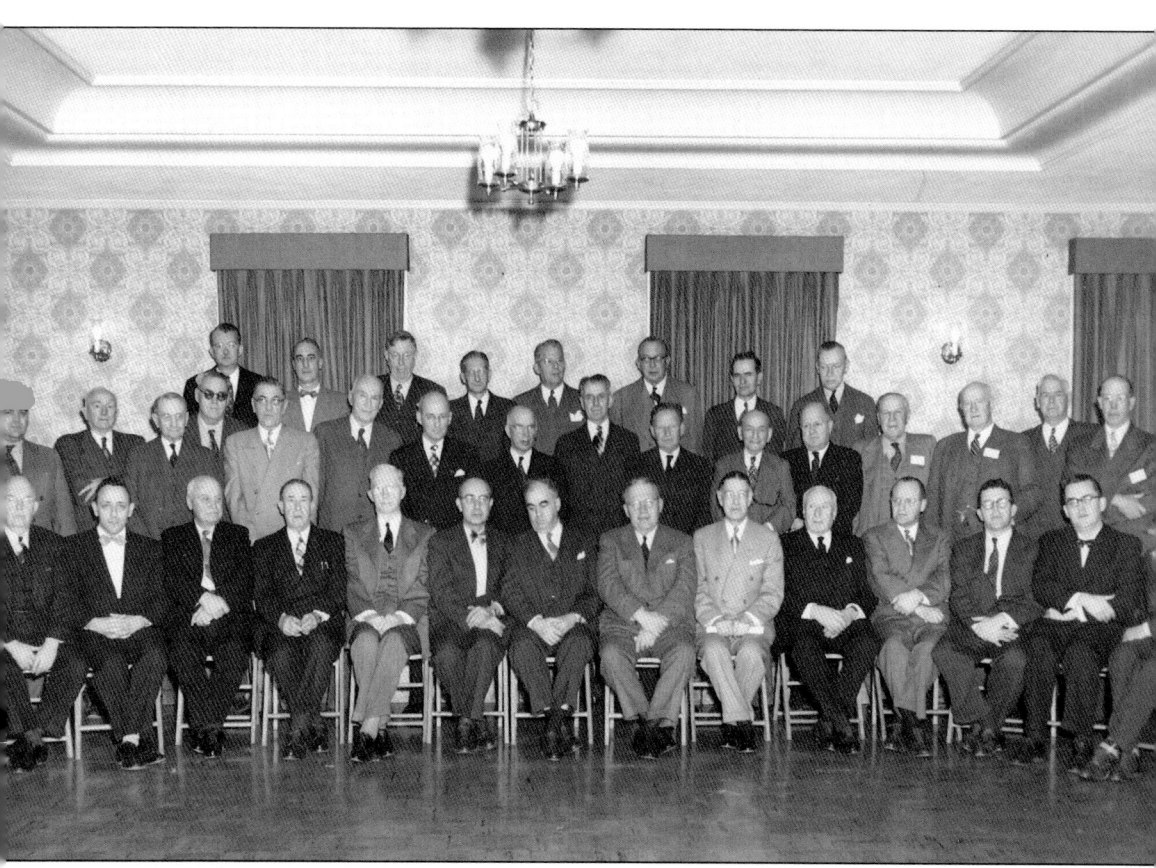

Present at the exposition's annual meeting on February 2, 1953, were the following trustees, from left to right: (first row) George L. Moore, John L. Rego, Harry L. Garrigus, Albert C. White Jr., Col. Frederick H. Payne, J. Loring Brooks Jr., Howard W. Selby, R. DeWitt Mallary, Maj. A. Erlan Goyette, Dudley P. Rogers, Frank P. Philbrick, Laurence R. Wallace, Jack Reynolds, and Wallace V. Camp; (second row) Russell Cushman, Edward H. Thomson, Horace A. Noble, Thomas A. Pearson, Edwin F. Weber, Fred E. Carlisle, F. Nelson Bridgham, Harley B. Goodrich, William B. Pape, Francis S. Murphy, Charles A. Frazer, Leonard J. Presson, Robert J. Cleeland, Henry P. Kimball, John Chandler, and Howard S. Rich; (third row) Raymond A. Loring, Dean Wilfred B. Young, William N. Howard, Charles E. Weeman, Ernest J. Wheeler, Raymond S. Redfield, Philip N. Darling, and Edward Ellingwood. At the meeting but not pictured were L.J. Kalmbach, Frances Osborne Kellogg, Saul M. Silverstein, and Willard B. Rogers. Howard Selby, president of the exposition since 1947, died unexpectedly just a month before the September opening of the 1953 fair.

Eighteen-year-old Fred Scoralick of Poughquag, New York, won the 1953 4-H Baby Beef competition with his 900-pound Red Angus Ferncliffe Laddie. On September 21, Pres. Dwight D. Eisenhower, while visiting the exposition, presented the youth with his grand champion banner. The president remarked on the rarity of the Red Angus steer, which greatly impressed Scoralick. The steer was auctioned for $1,450; First National Stores was the high bidder.

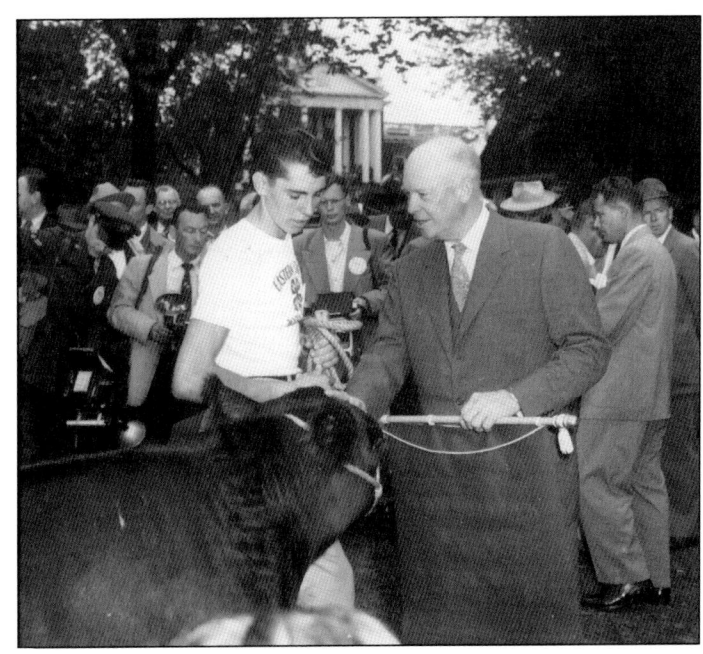

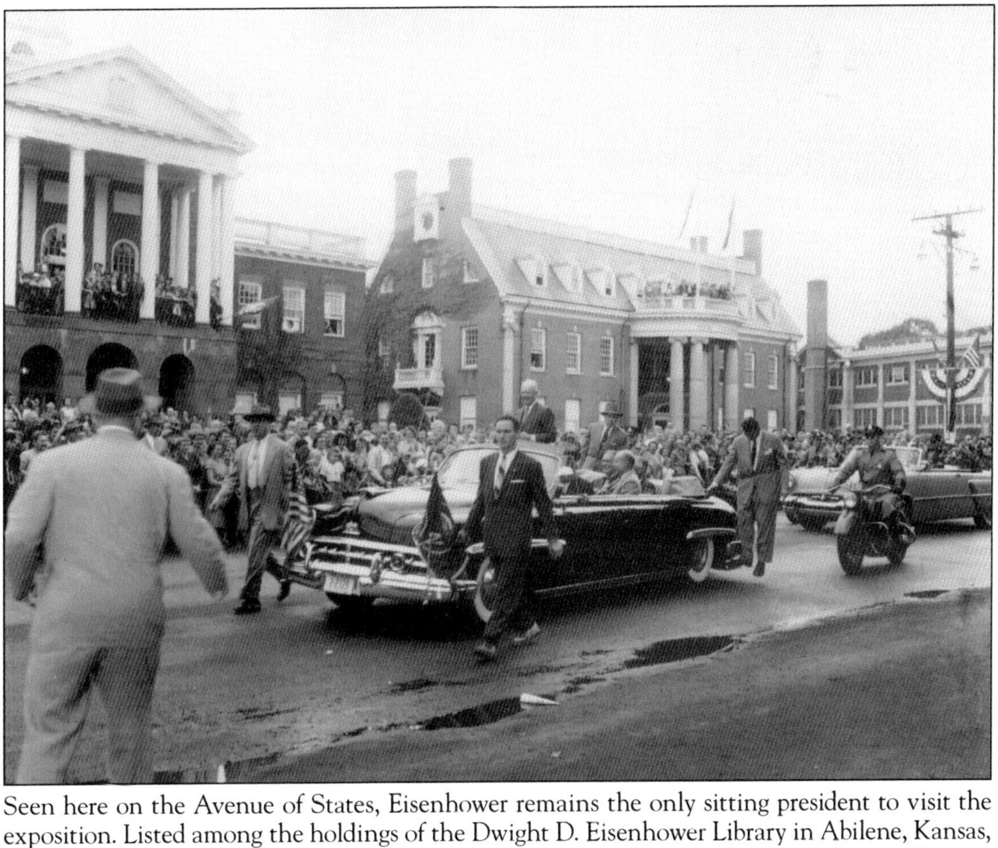

Seen here on the Avenue of States, Eisenhower remains the only sitting president to visit the exposition. Listed among the holdings of the Dwight D. Eisenhower Library in Abilene, Kansas, are "motorcade and walking route; Storrowton: the New England Village; 31st Annual Eastern States Exposition," and "Eastern States Exposition maps, buttons and credentials."

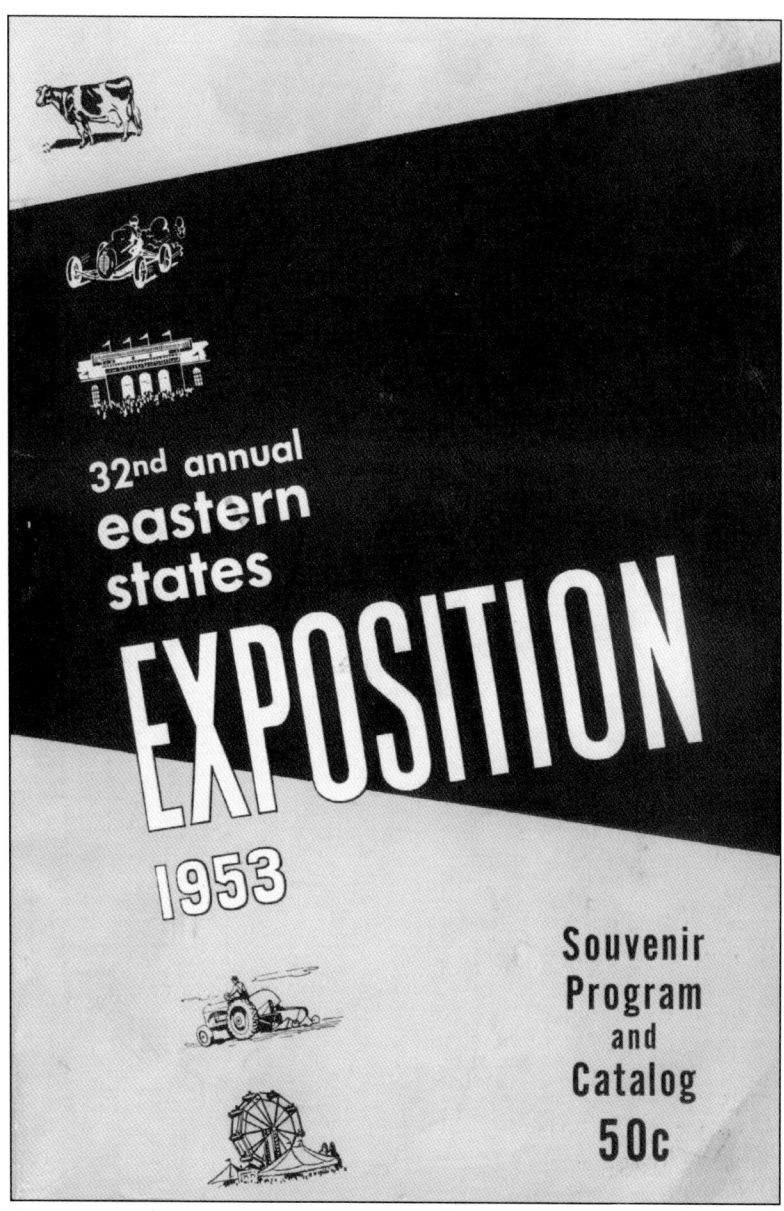

The exposition program has always been a useful tool for visitors, containing detailed descriptions and schedules of events, along with exhibitor and concession information. For more than three decades, entrant information for all livestock contests was also included. As interest in the exposition continued to grow, so did the program; the 1952 edition filled 256 pages, 127 of which were livestock entries. A slimmer 1953 program (128 pages) eliminated the livestock entries, which were printed in separate publications. The fair's permanent buildings and features, new attractions, and participating youth groups were the focus of that year's program. A Fireman's Muster with $1,000 in prize money was organized for 1953, and the internationally acclaimed water show Dancing Waters was held in a specially constructed tent outside the Industrial Arts Building. The practice of using cover designs for several years was also discontinued during the early years of the 1950s; going forward, the cover of the program would feature a new design each year. The last year for the familiar six-inch-by-nine-inch format was 1954. (Author's collection.)

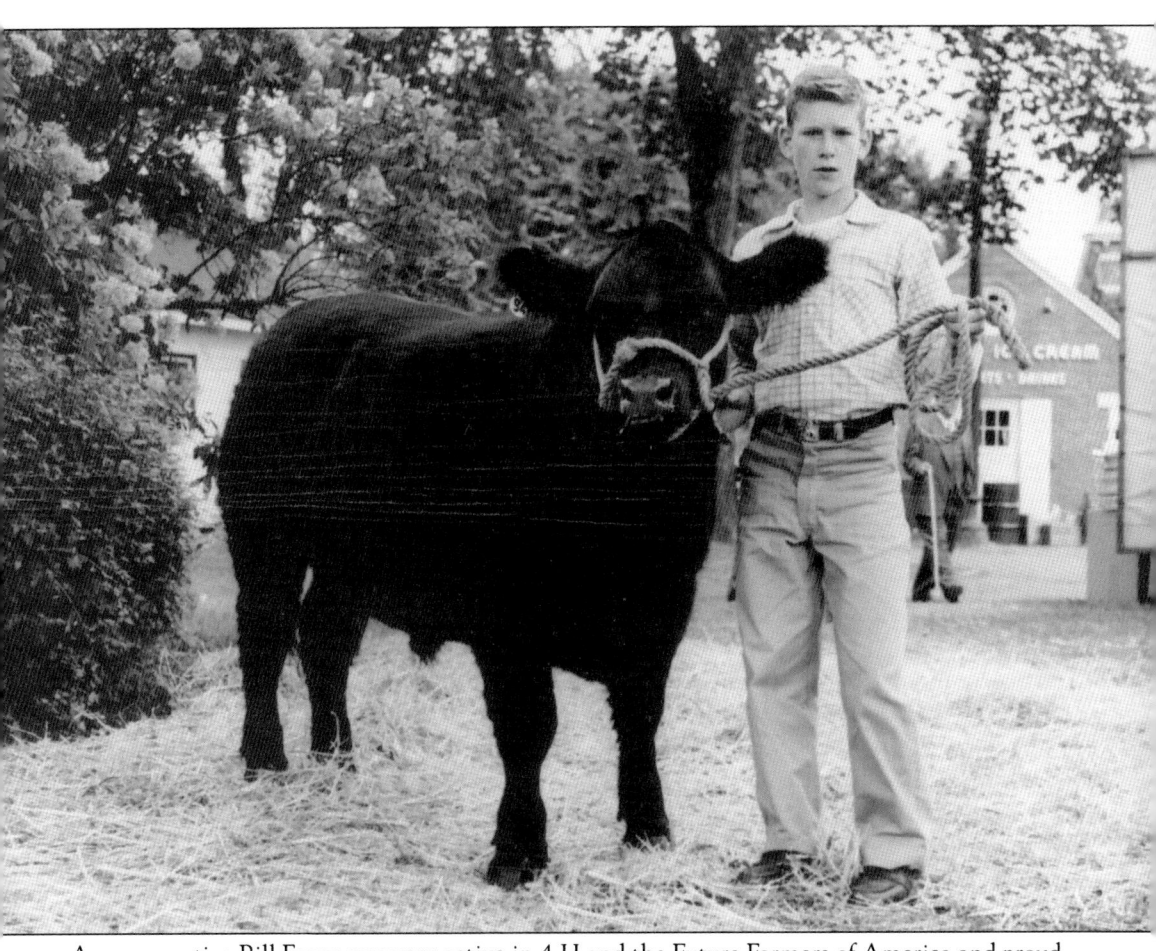

Agawam native Bill Fearn was very active in 4-H and the Future Farmers of America and proud of it. Seen here with his Angus Bywood Blackie III in 1956, he was awarded the Baby Beef Blue Ribbon and the Heavyweight Steer Award at the 4-H Fair held that August on the exposition grounds. The previous year, Fearn's entry was Bywood Blackie II, purchased as a 370-pound calf from Bywood Farm in Canaan, Connecticut, for $81.40. Baby Beef participants were required to keep detailed expense records; added to the purchase price was $16.90 (roughage), $162.17 (feed), $2.50 (veterinary expenses), $6 (insurance), and $3 (entry fee), for a total cost of $271.97 to raise the calf. A detailed narrative of the steer's care and feeding was also required. Fearn described the steer's eating habits, the automatic feeder he constructed, and repairs necessary to Blackie's pen in the wake of Hurricane Diane. Fearn would later show and judge livestock across New England and New York. A gracious and generous gentleman, he still recalls fondly his long association with the 4-H and the exposition. (Courtesy of Bill Fearn.)

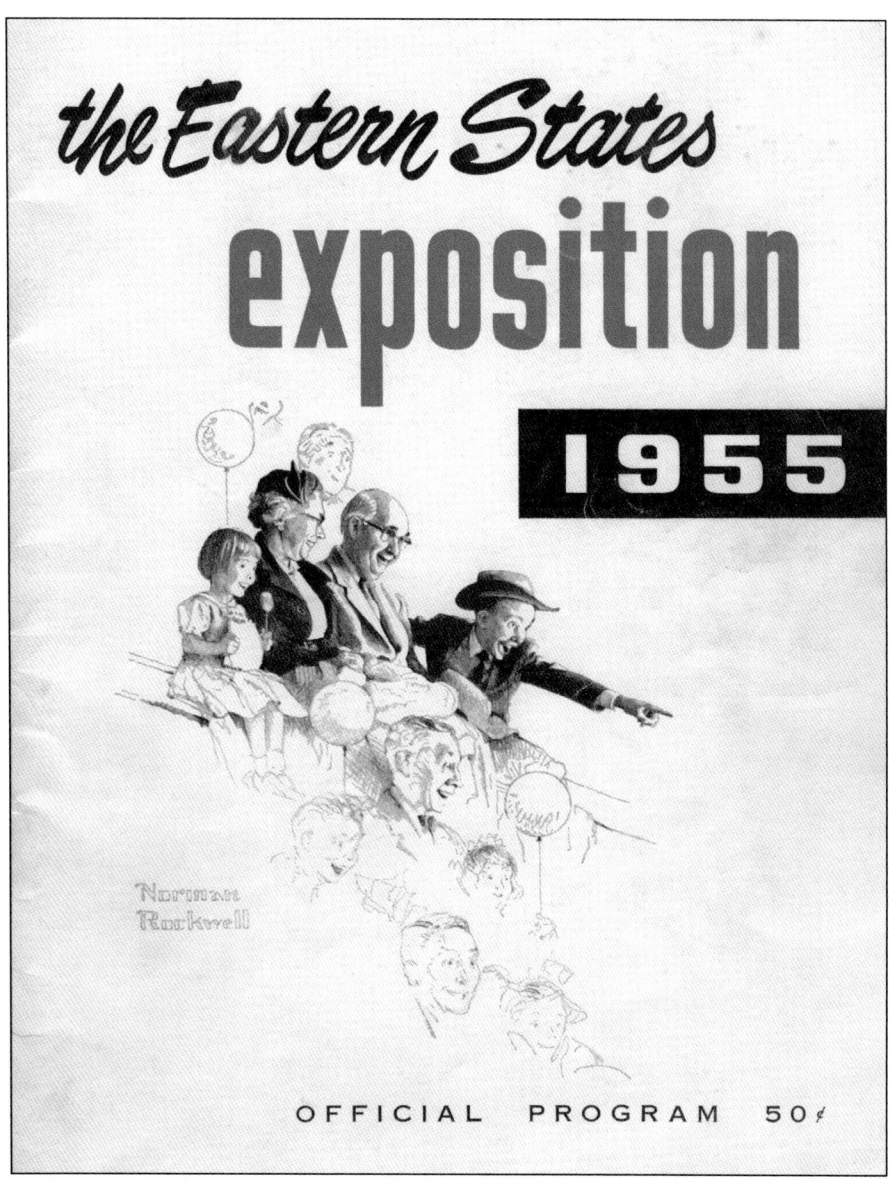

A ninth day was added to the exposition's run in 1954; a new, larger 8.5-inch-by-11-inch program booklet debuted the following year. The cover that year featured an original drawing by Norman Rockwell that was made available to the exposition by the Massachusetts Mutual Life Insurance Company of Springfield. As detailed in that year's program, this illustration was chosen because fair officials felt that it "best personified the family atmosphere which prevails at the Exposition." Contributing to that family atmosphere were the JE Ranch Rodeo, a huge display of more than 600 varieties of tropical fish put on by the Pioneer Valley Aquarium Society, the New England Open Drum and Bugle Corps competition, the Mummers Four Leaf Clover String Band, an appearance by Les Paul and Mary Ford, Jack Joyce's Performing Camels, the Country Music Time show, and the Ice-A-Rama "stage show on ice." The year 1955 also saw, for the first time ever, a parade held on the grounds. Stepping off at 1:00 p.m. daily, the mile-long parade included a Festival of Floats depicting "various phases of history and economy in the New England States." (Author's collection.)

After the business meeting at the New England Press Photographers Association's annual convention at the exposition in May 1955, the more than 50 voting members held a beauty contest for the twin titles of Miss New England Press Photographer and Miss Exposition. Beverly Jansen of Providence, Rhode Island, far right, was selected from among 20 contestants. She was described in the 1955 exposition program as "beauteous" and "pretty, pert, and photogenic" by the *Springfield Republican*.

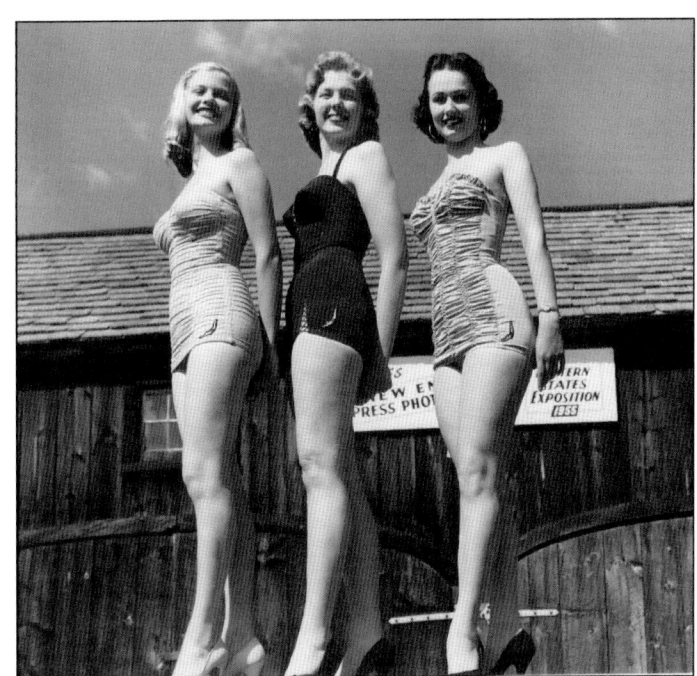

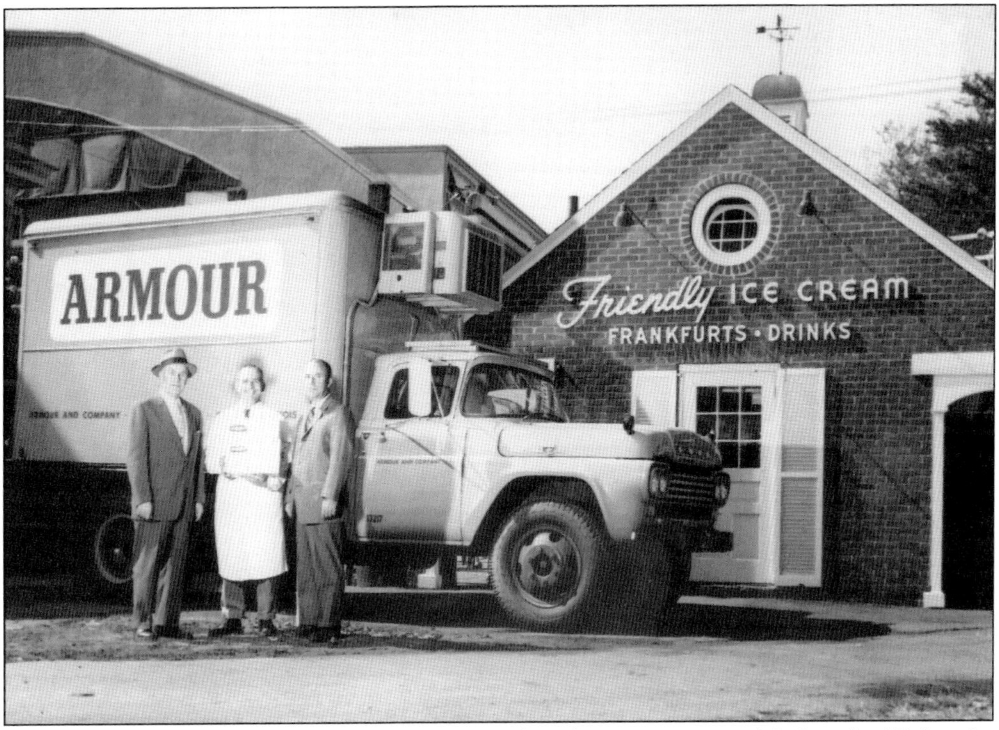

Agawam native Eddie Neilson delivers Armour hot dogs to Curtis and S. Prestley Blake, the founders of Friendly Ice Cream, outside the Friendly concession near the Coliseum around 1955. Want ads that fall read "Neat appearing men for full time work with Friendly Ice Cream at the Exposition. Good Pay. Apply in person at Friendly Ice Cream Booth at the Exposition." (Courtesy of Nancy Neilson.)

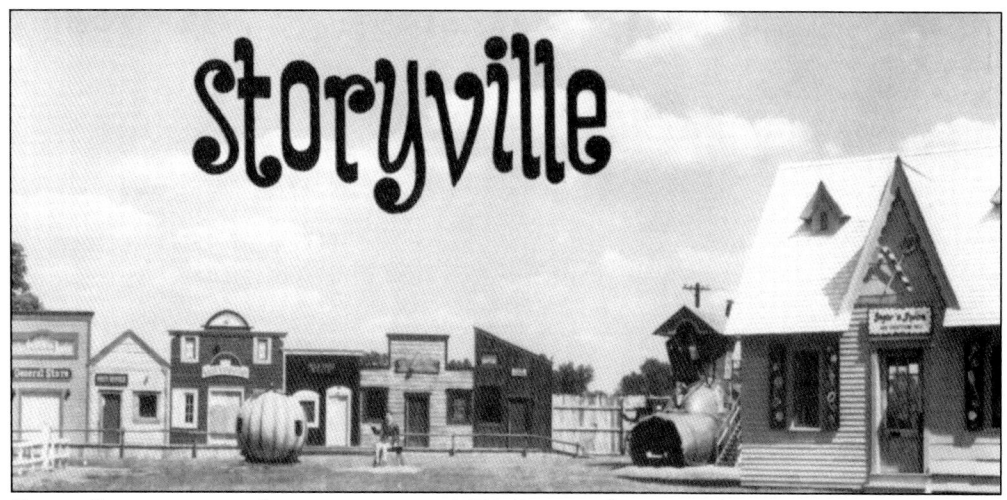

Storyville was new at the exposition in 1956. Designed to appeal to younger fairgoers, the "enchanting land of make-believe" featured fairy tale and Mother Goose characters, and a Frontier City. Moby Dick spouted water 10 feet into the air. The Old Woman who Lived in a Shoe feature was very popular; children could slide down and out the toe of the shoe on a specially built chute. (Author's collection.)

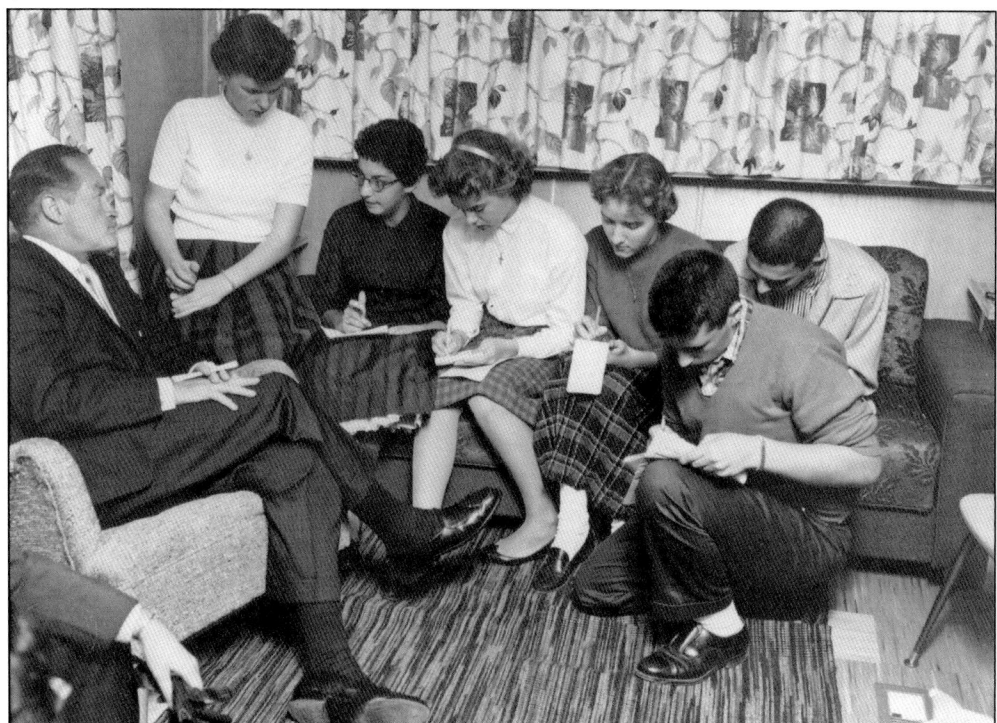

Bob Hope appeared in the Coliseum each evening at 8:00 p.m. from September 15 to 21, 1957. Stars of Hope's stature are always newsworthy, and getting the scoop from the man himself for their high school newspapers are, from left to right, Wanda McNeely of St. Mary's High School in Westfield, Massachusetts; Joyce Jadelson, Holyoke (Massachusetts) Catholic; Isabel Millane, Holyoke High School; Frances Balch, St. Mary's; and Alan Mathewson and Ronald Desnoyers of Chicopee (Massachusetts) High School.

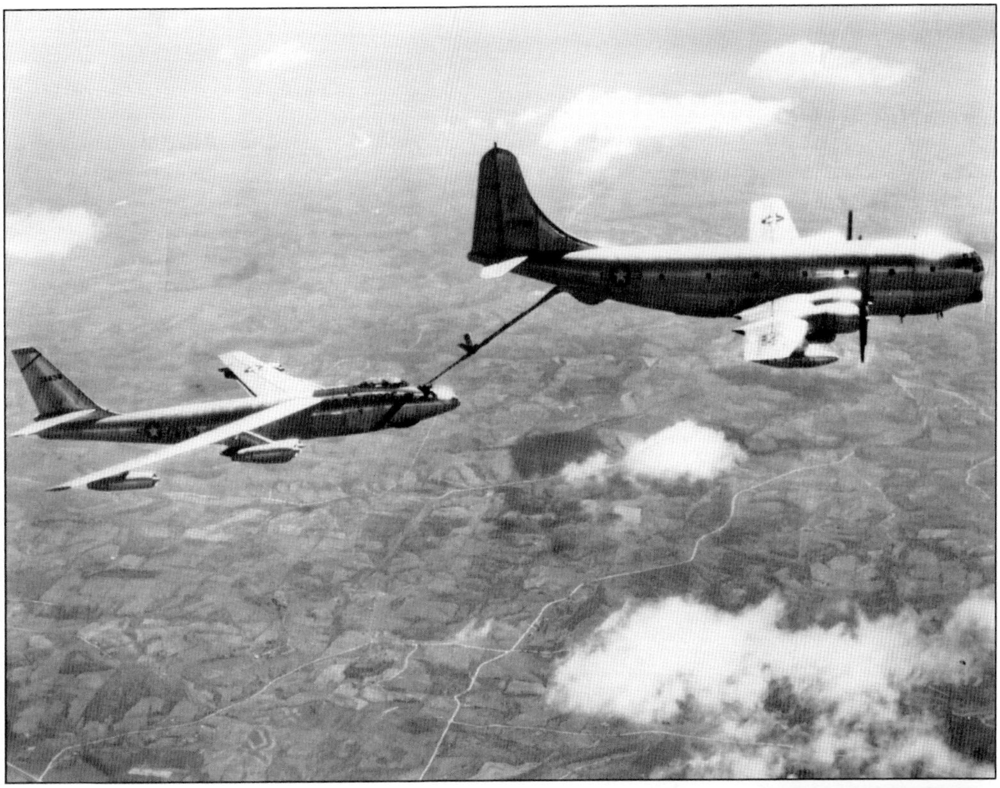

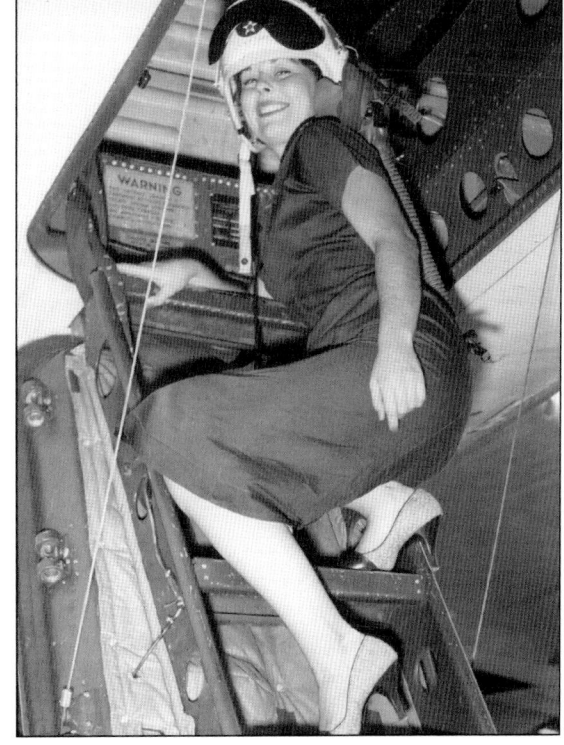

A B-47 bomber is refueled by a KC-97 tanker in the skies over New England during the exposition's 1957 observance of the US Air Force's 10th anniversary. Featured were exhibits throughout the fair's buildings and grounds, performances by Air Force musical ensembles, an air show by the USAF Thunderbirds, and a visit from Air Force chief of staff Gen. Thomas D. White. Among the most popular attractions was a B-29 bomber fuselage. (Author's collection.)

Miss Exposition 1957, Joyce Cunliffe of Manchester, Connecticut, is seen here on the entry ladder of a Boeing B-52 Stratofortress at Westover Air Force Base in Chicopee, Massachusetts. The frontline Strategic Air Command bomber, B-47 Stratojet and B-66 Destroyer bombers, F-94 Starfire interceptors, F-86D Sabrejet and F100 Super Sabre fighters, along with other aircraft, took part in an air show over the exposition grounds.

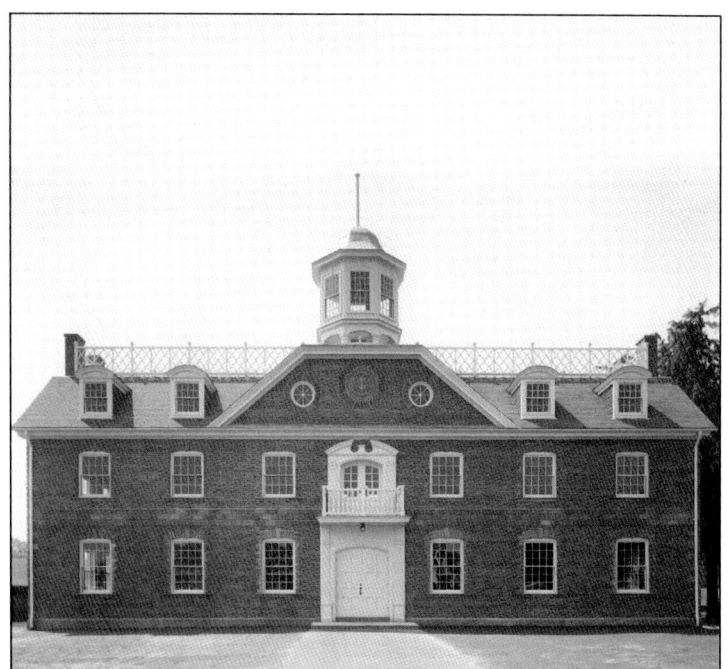

The long-awaited Rhode Island State Building was finally dedicated on September 21, 1957, completing the Avenue of States project that began nearly 40 years prior. More than a dozen paintings by Henry Farre, part of a collection given to the new US Air Force Academy by Laurence Rockefeller, were on display in the new state building's Air Force Lounge.

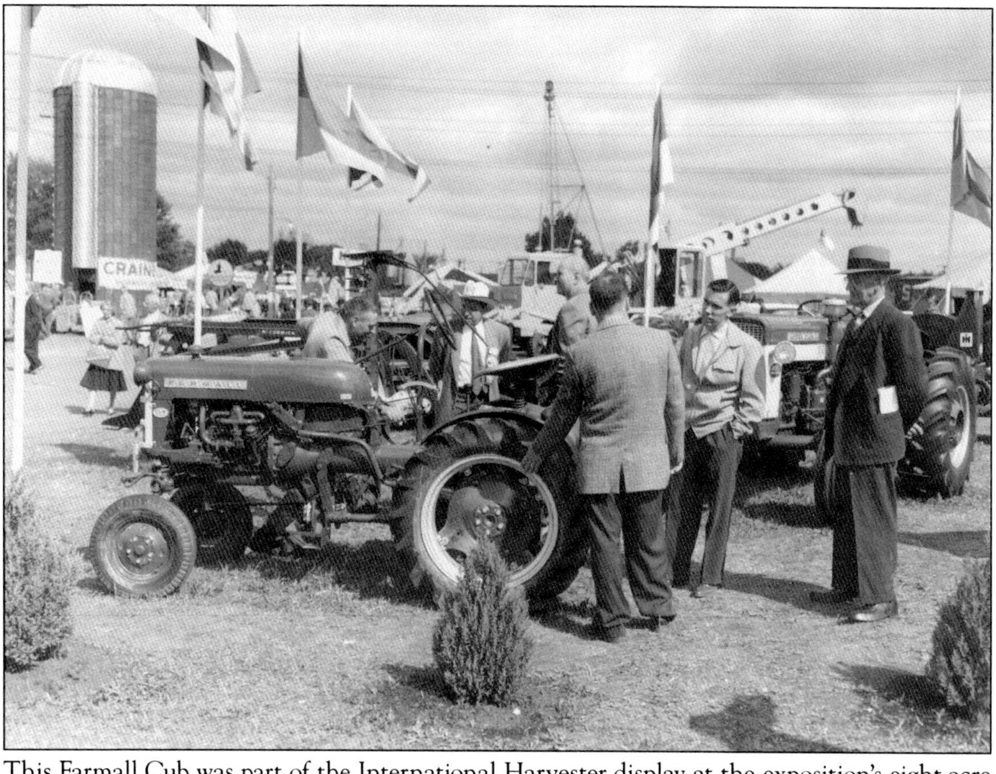

This Farmall Cub was part of the International Harvester display at the exposition's eight-acre outdoor farm machinery show. The versatile and nimble tractor was extremely popular on small farms; a large number of the machines are still in use to this day. The International 460 utility tractor seen at right was introduced in 1958 and manufactured until 1963.

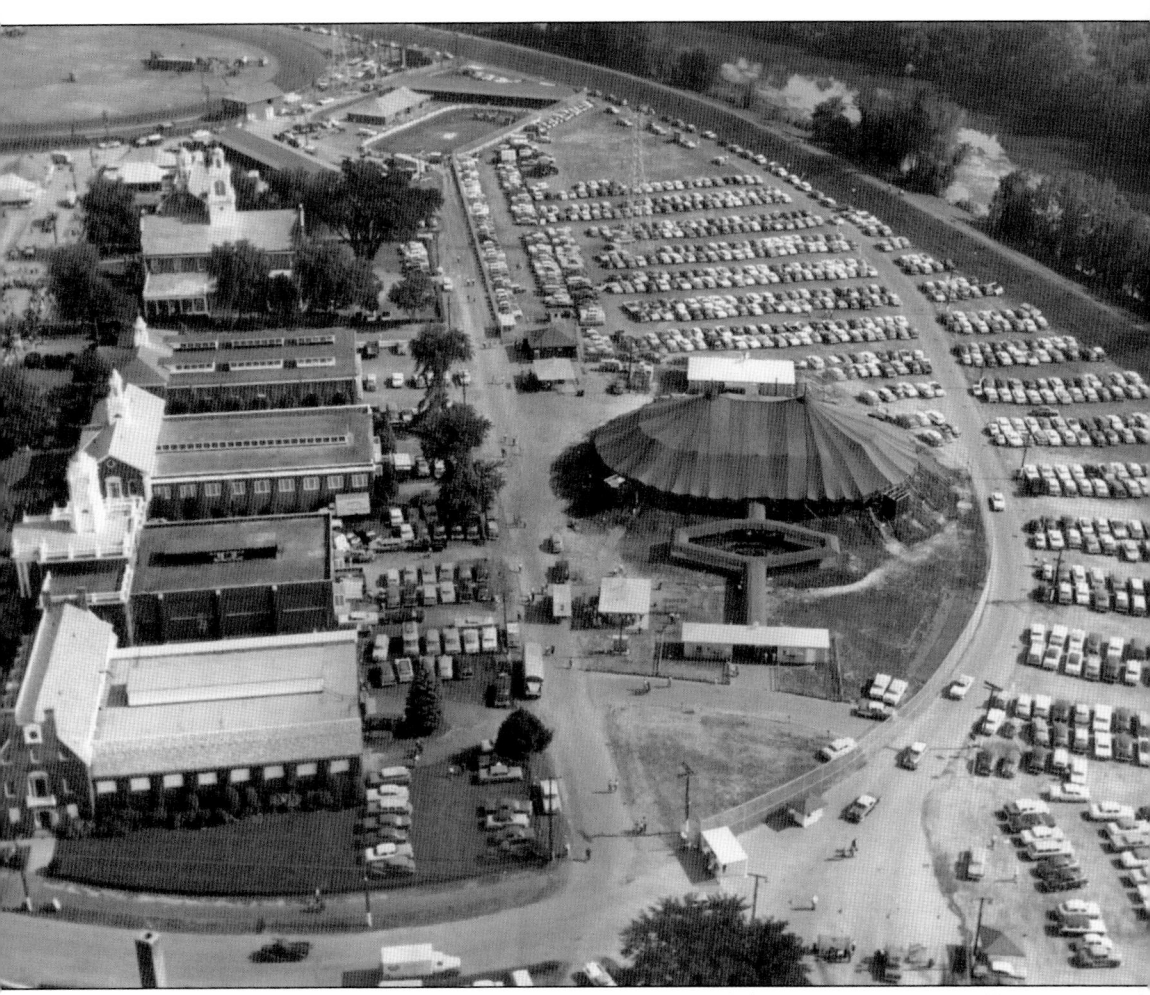

A new era of regional live theater began with the opening of the Storrowton Music Fair on June 15, 1959. Productions were staged in a large, oval tent located west of the exposition's state buildings. The gala premiere season began with *The King and I* and ended on September 5 with *Oklahoma!* During Exposition Week that year, the tent was utilized for performances by the Ballet Español Ximenez-Vargas, comedian Herb Shriner, and a jazz festival featuring Woody Herman and the Chris Barber Jazz Band. Wally Beach ran the summer theater program as Storrowton Music Fair until 1967. That year, the theater was purchased by Mike Iannucci and Ann Corio, who operated it as Storrowton Theater until the stage went dark in 1978. In the two decades it was in operation, a who's-who of musicians, entertainers, and actors appeared in the tent. In 1985, the *Sunday Republican*'s "artful codger," R.C. Hammerich, lamented, "Summers have never been the same around here since the Storrowton Theater on the Eastern States Exposition grounds closed." Thirty years later, that is still true.

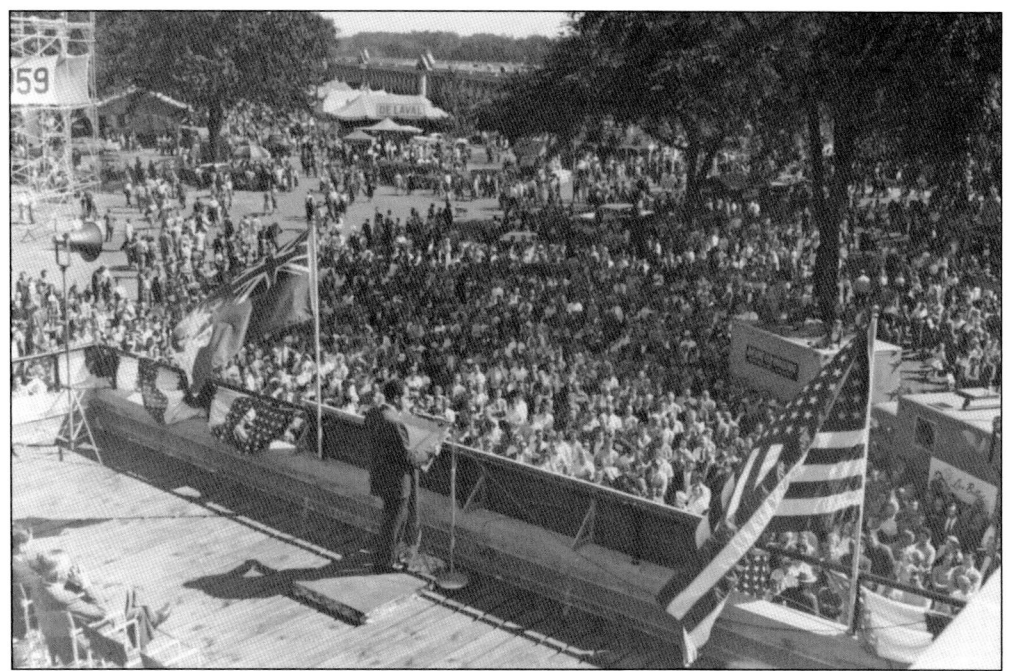

A perfect autumn day, an appearance by Guy Williams (better known as television's Zorro), and a visit by Vice Pres. Richard M. Nixon all combined to create massive traffic jams leading to the exposition and a record crowd of 91,347 visitors on Sunday, September 20, 1959. That morning, the vice president toured the state buildings and spoke at a luncheon with exposition trustees and New England political and business leaders. Later that day, Nixon spoke from the terrace of the Brooks Building, telling a crowd of approximately 15,000 that the exposition represented "America at its best." During a visit to the third annual Eastern States Art Exhibit, co-sponsored by the exposition and Springfield's Museum of Fine Arts, Nixon was presented with the watercolor *Snow in the Foothills*, painted by Wilbraham, Massachusetts, resident Richard C. Stevens. (Below, author's collection.)

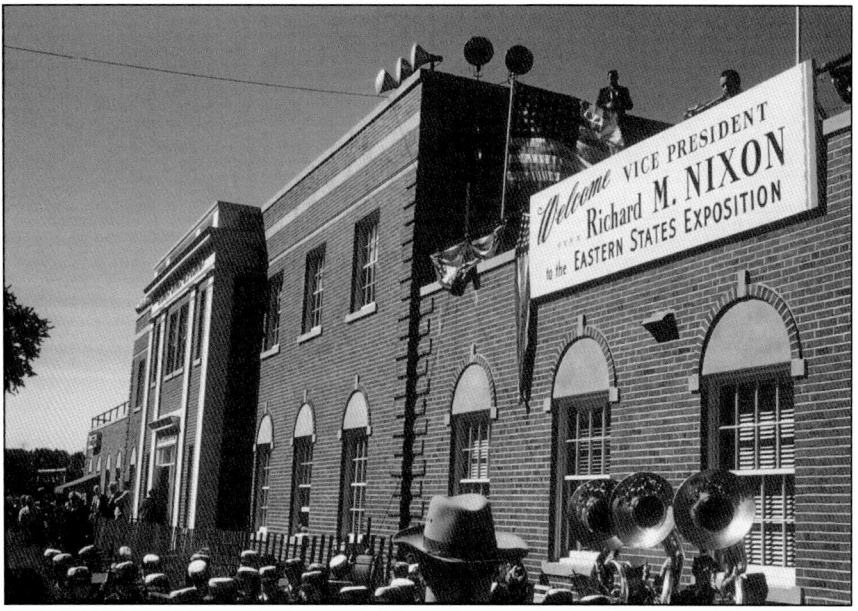

Feeding Hills native John Janik started at the exposition as a groundskeeper in the early 1950s, working with grounds superintendent Edward Plante. He later became Plante's assistant and was named the exposition's superintendent of the Grounds Department in 1967. He retired in that capacity in 1972. He is seen here in the early 1960s moving one of the exposition's tram cars. (Courtesy of Ellen Janik.)

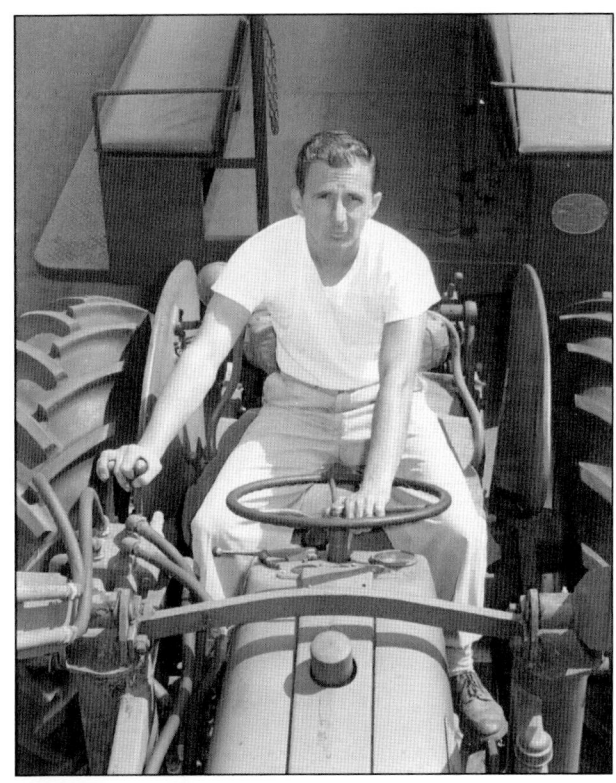

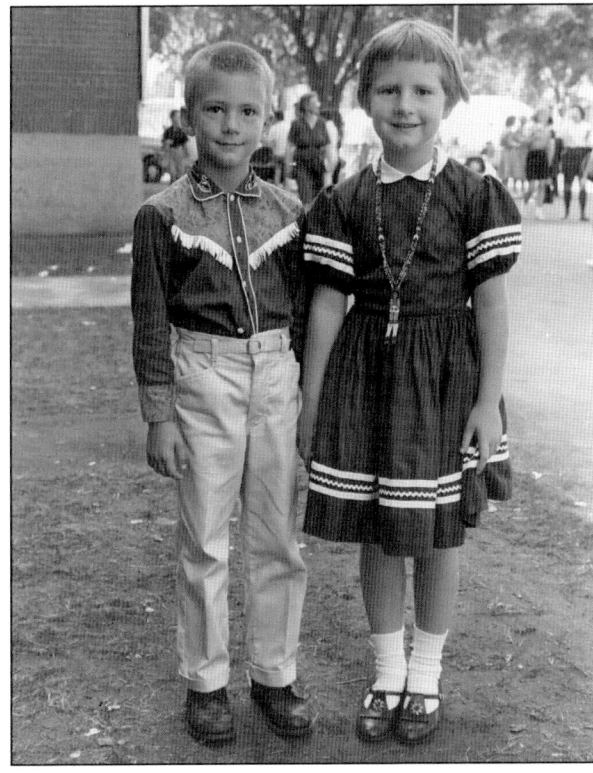

Janik's children, Doug and Cindy, seen here around 1961, understandably spent a lot of time at the fair. Among that year's attractions were an air show by the US Navy Blue Angels; Jack and Jill, the laughing porpoises from Miami's Seaquarium, in their specially constructed Aqua-Dome; a replica of a Mercury space capsule provided by NASA; and examples of the US Army Air Defense Command's Nike missile family. (Courtesy of Ellen Janik.)

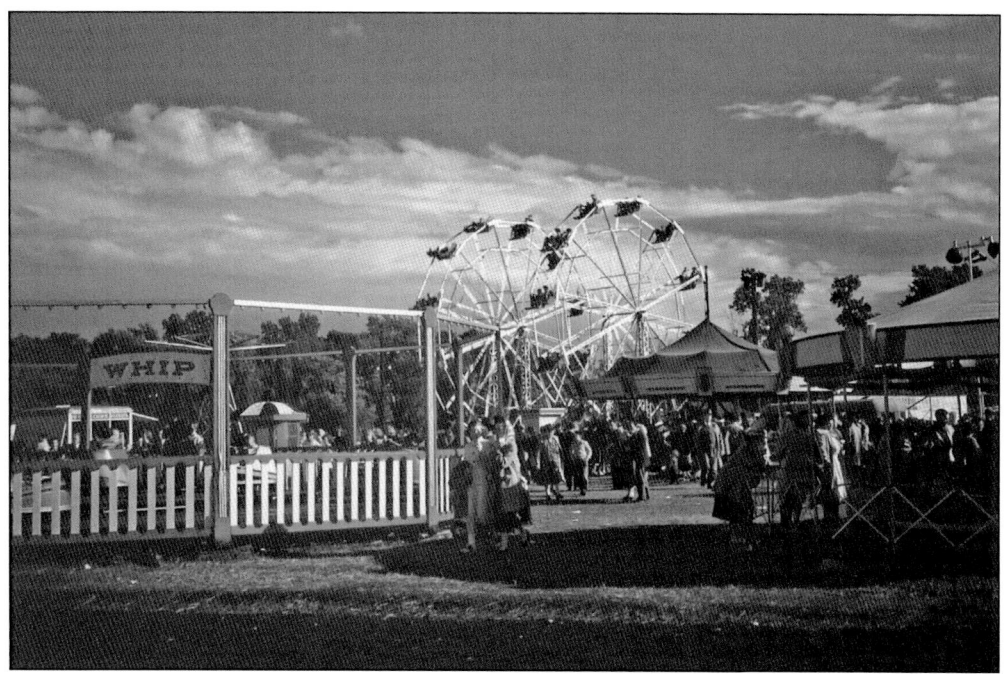

The September 25, 1961, *Springfield Union's* "Exposition Sidebars" noted that the six doctors and two nurses at the fair's hospital treated nearly 2,000 patients each day. "Fathers who were induced into riding the whip and the Ferris wheel with their youngsters made frequent patients, the nurses said. The kids made out fine on the rides, but sometimes their fathers would come in here with upset stomachs." (Author's collection.)

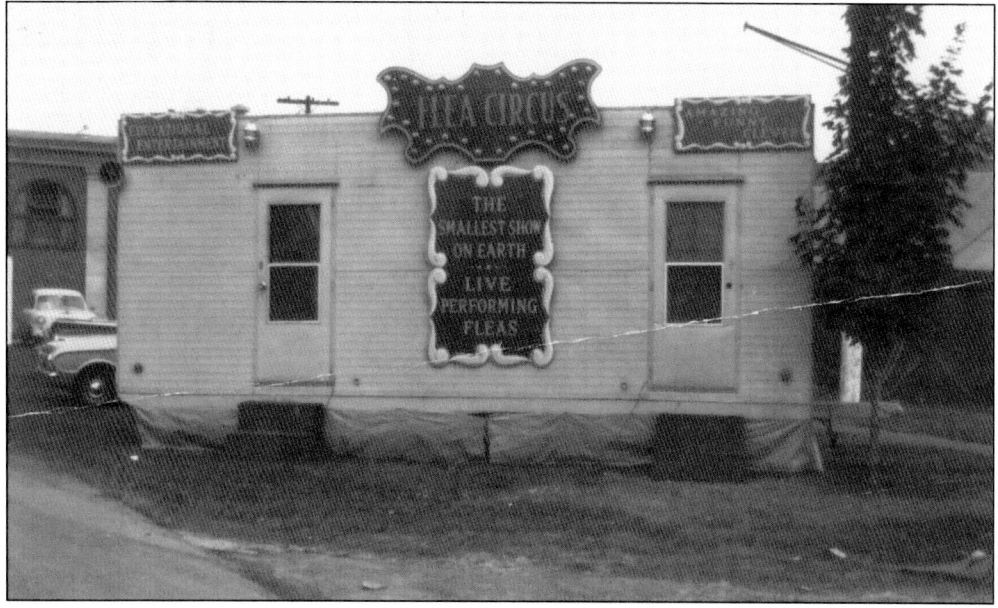

One of the more unusual "livestock" exhibits of 1961 was housed in this small trailer behind the horse barn. John Torp's flea circus, one of only four in the world, featured 10 fleas performing in a five-minute show. The fleas were trained to run, jump, and juggle; it was necessary to view the one-ring circus through a magnifying glass.

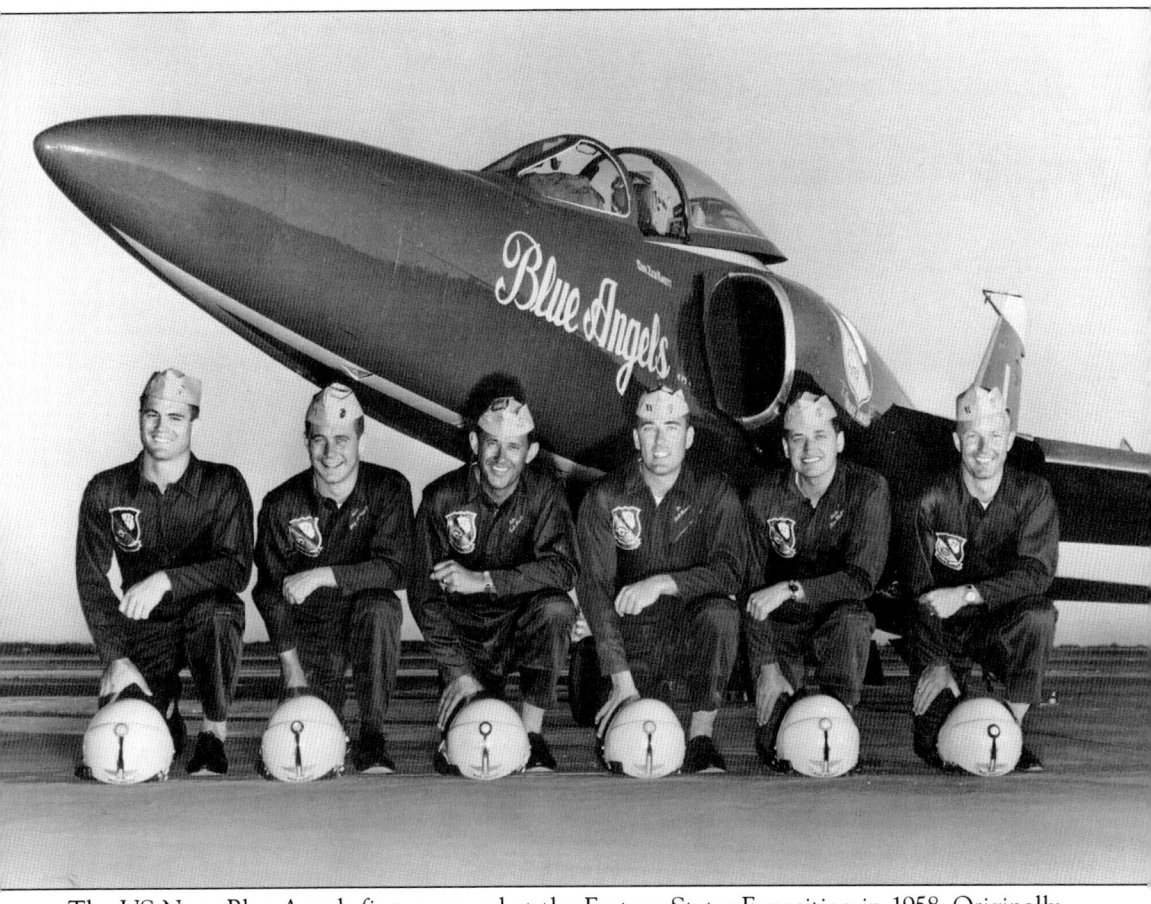

The US Navy Blue Angels first appeared at the Eastern States Exposition in 1958. Originally scheduled to appear September 17, 18, and 19, inclement weather caused the cancellation of their first two shows, plus that of another that was added on September 20. The only flight that year was on September 19, abbreviated due to a jet fighter scramble at Westover Air Force Base that delayed takeoff and limited Civil Aeronautics Authority clearance in local skies. The Blue Angels fared better during their 1959 return; perfect weather resulted in record crowds, including members of the Royal Canadian Air Force's Golden Hawks demonstration team, who had performed earlier in the week and who were able to watch the show from the roof of the Coliseum. The Blue Angels' final appearance at the exposition was in 1961, when they again enjoyed blue skies. The USAF Thunderbirds performed in 1957, 1960, and 1962. Air shows over the exposition were later discontinued, attributed to increased red tape in obtaining permits for shows over populated areas.

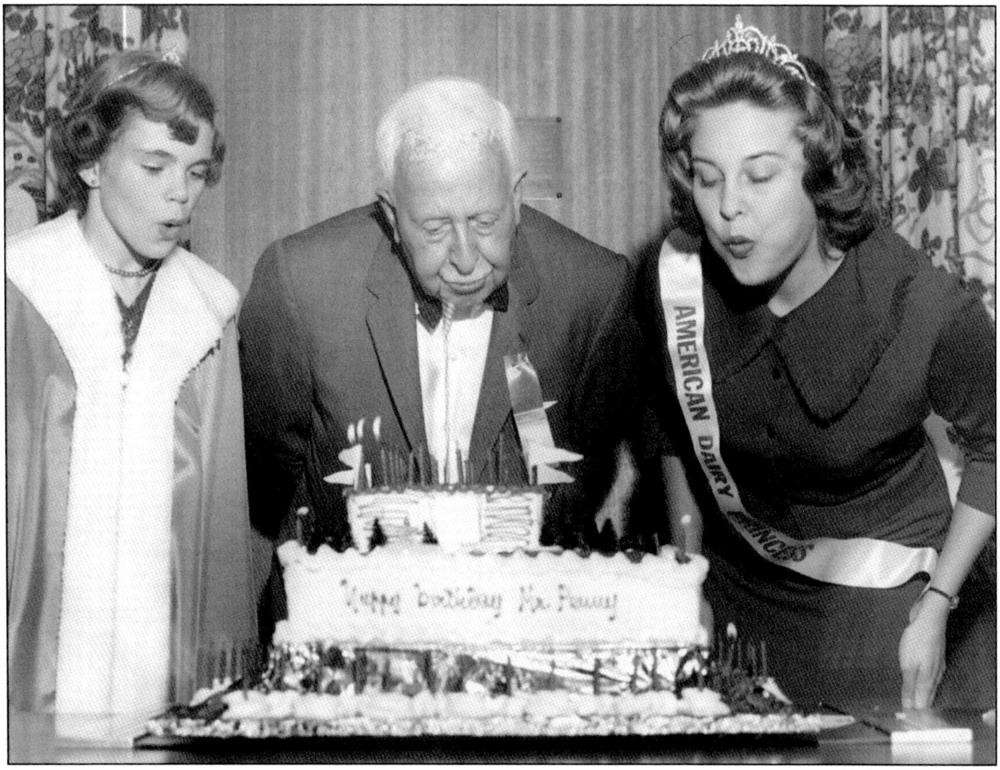

The 1961 Eastern States Exposition opened on September 16, the 86th birthday of longtime exposition trustee James Cash (J.C.) Penney, of department store renown. Shown here blowing out candles on his birthday cake are, from left to right, Exposition Princess Susan Bunnell, Penney, and 1962 American Dairy Princess Louise Knolle. In 1964, Penney was presented with a proclamation signed by the six New England governors in recognition of his 37th year as a trustee.

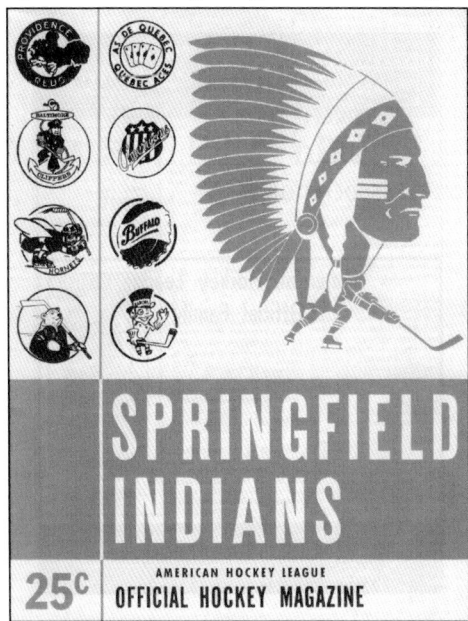

The Springfield Indians, one of the original teams of the American Hockey League, share a long history with the exposition. The Coliseum was the Indians' home ice from 1926 to 1972 and 1976 to 1980. The team played at the Springfield Civic Center from 1973 to 1975 and again from 1981 to 1994. Boston Bruins defenseman Eddie Shore, elected to the Hockey Hall of Fame in 1947, owned the Indians from 1939 to 1976. (Courtesy of Michael Cecchi.)

Four
BECOMING THE BIG E

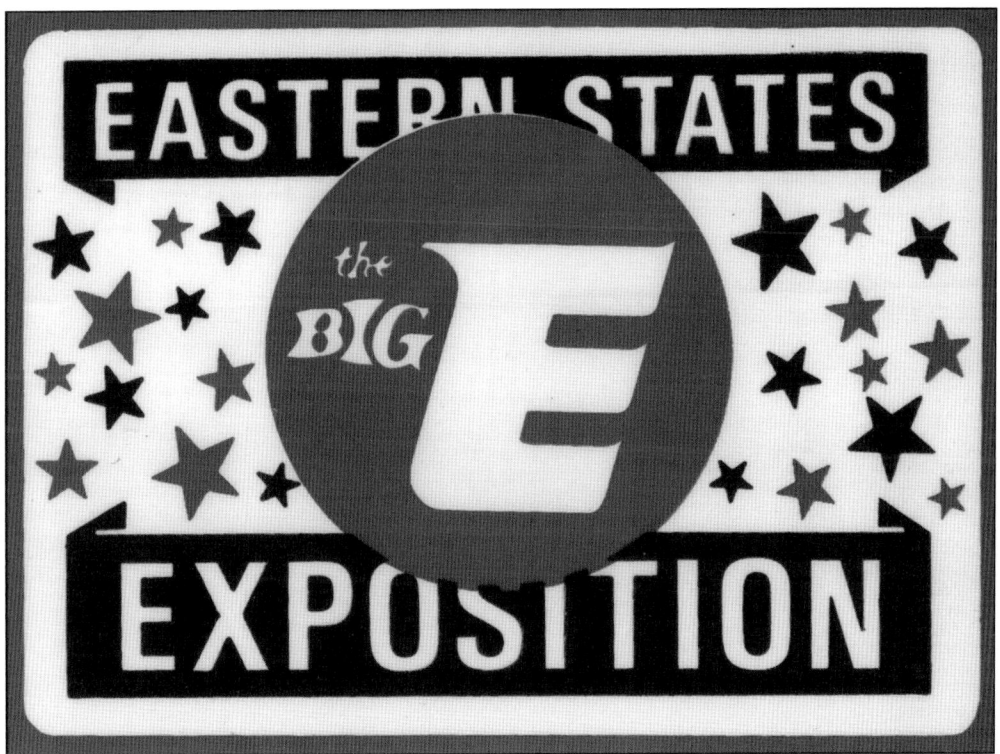

Remington Advertising developed the moniker "The Big E," which first appeared on the cover and contents page of the 1966 program; the following year, the cover and page headings utilized the name, though text still referred to the fair as "the Exposition." It was not until 1968 that general manager G.W. Wynne used "The Big E" in his introduction letter and the term was used throughout the booklet. It was Education, Excitement, and Entertainment for Everyone!

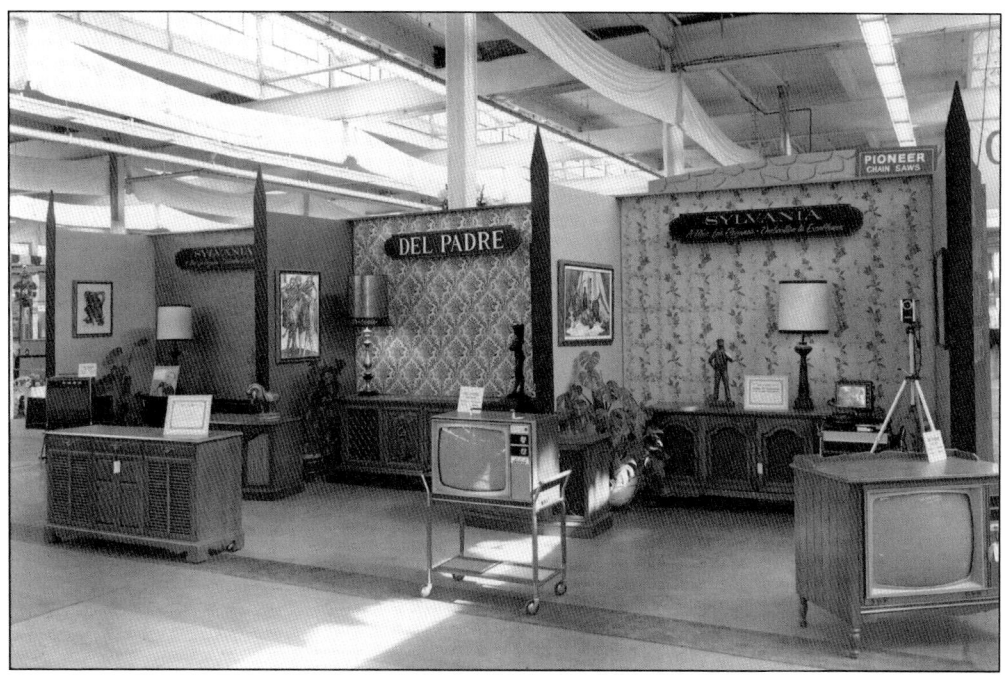

It was in 1957 that the Industrial Arts Building was first billed as the home of "Better Products for Better Living." In 1961, the building was renamed the Better Living Center. In the 1960s, better living meant a new television or stereo. For decades, the Del Padre stores, operated by Louis Del Padre, offered home electronics to western Massachusetts residents.

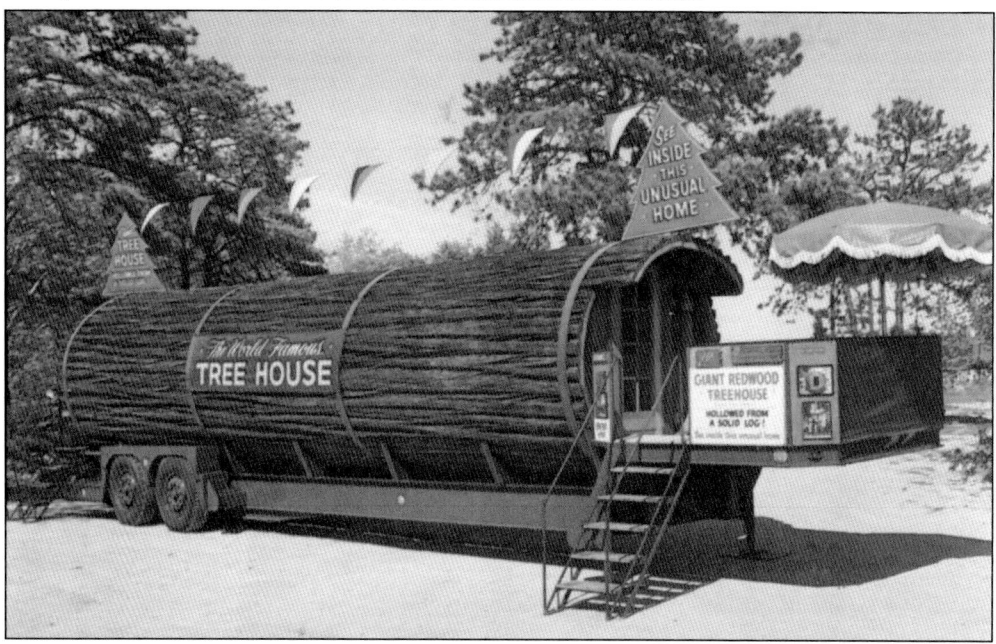

The Giant Redwood Treehouse was a popular attraction at the fair during the late 1960s and 1970s. It took Maxwell Kitson of Brigantine, New Jersey, a year to hollow out the 100-foot-long, 33-foot-wide log and create the house. The 12-ton, 980-year-old log contained a living room, bedroom, kitchen, and bathroom. (Author's collection.)

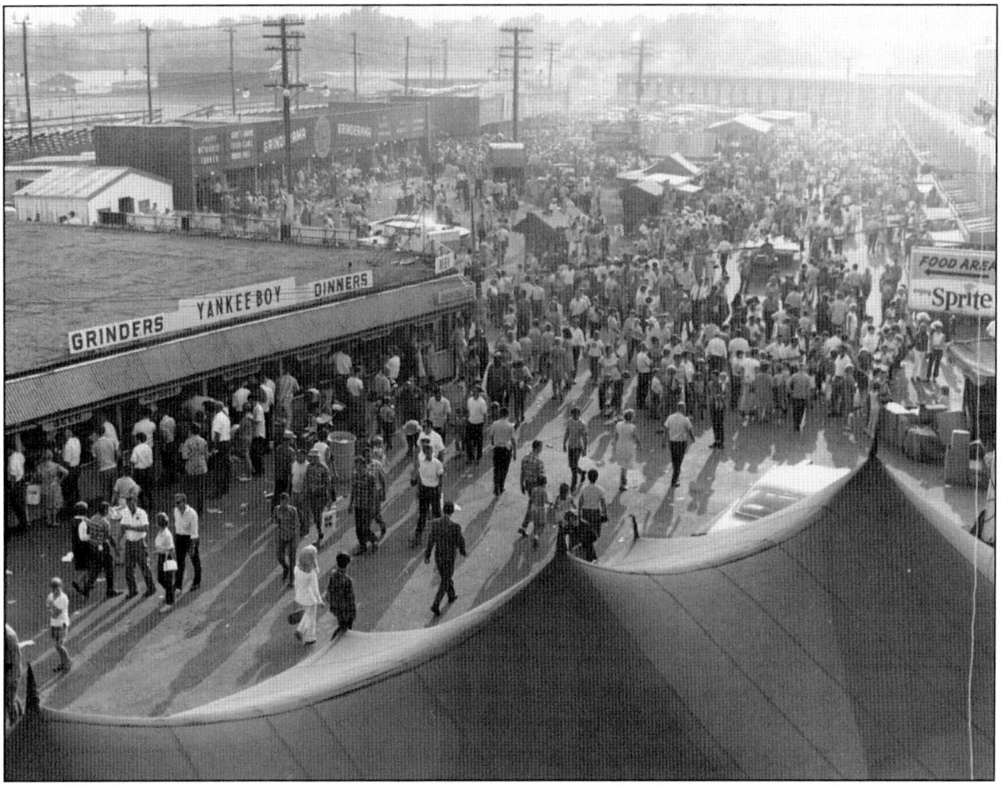

Located for many years behind the Coliseum were the popular Grinderama (left background) and Yankee Boy concessions. Anthony "Sharkey" and Jennie Dialessi operated Grinderama for nearly 50 years. The building was demolished in 1994 and is now the site of Kiddieland. Yankee Boy is currently located on the fair's Springfield Road and operated by Glen Bouchard, grandson of founder Peter Ortolani, who started at the fair selling hot dogs in 1926.

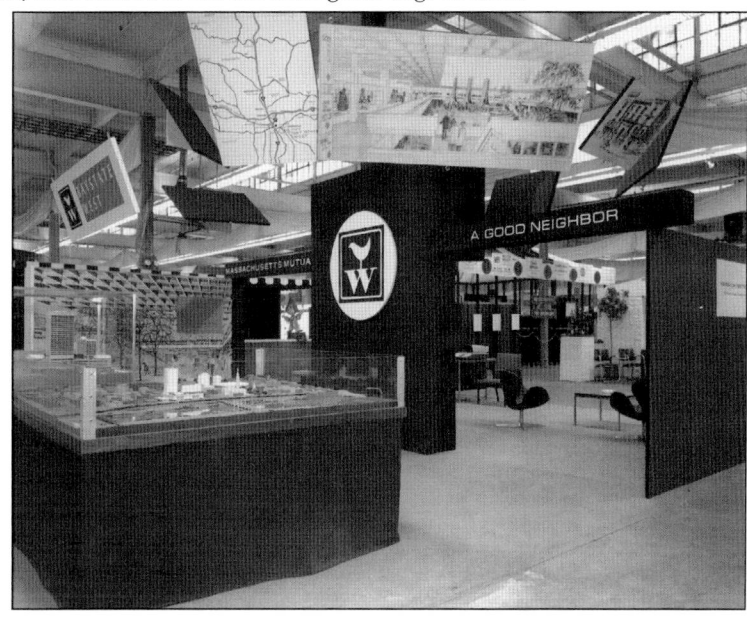

In 1967, Massachusetts Mutual Life Insurance Company exhibited plans for its Baystate West development project. The display in the Better Living Center included a model of the buildings to be constructed in the heart of Springfield. Initially named New Springfield, the $50 million downtown revitalization project opened in 1971 and is currently known as Tower Square.

107

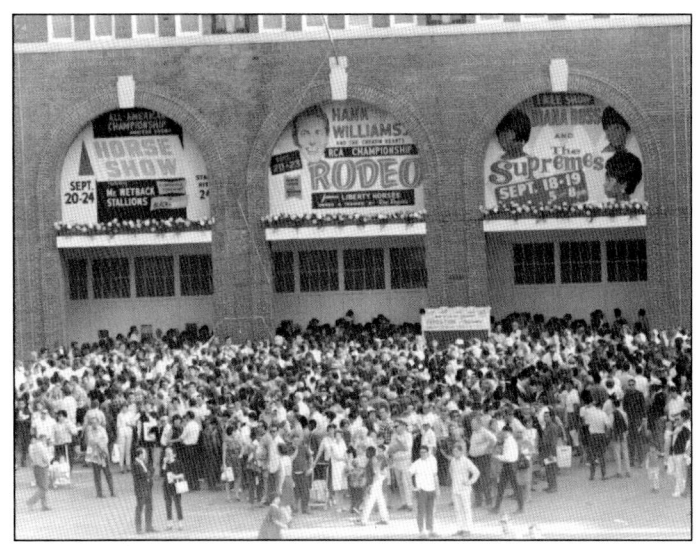

Diana Ross and the Supremes' "The Happening" and "Love is Here and Now You're Gone" had both hit number one on the charts in early 1967. Their appearance at the exposition on September 18 and 19 of that year ushered in free Coliseum shows and the debut of a $100,000 lighting and speaker system in the arena. The new 14-speaker sound system was a marked improvement over the previous setup.

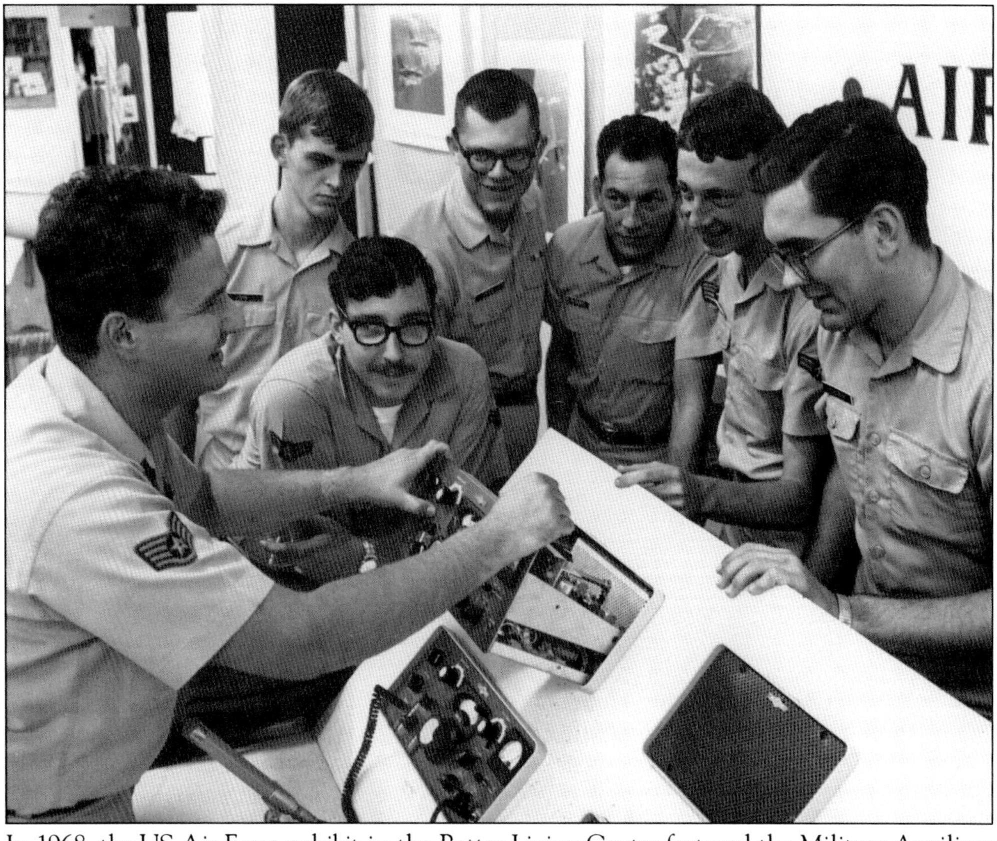

In 1968, the US Air Force exhibit in the Better Living Center featured the Military Auxiliary Radio System (MARS), allowing fairgoers to send messages to certain areas of Vietnam. Seen here are, from left to right, Sergeant Caldwell, A1c. Mike John, A1c. Nathan Pius, A1c. Lawrence Bielski, S.Sgt. Juan Palacios, A1c. William Burkhardt, and A1c. John Furgeson.

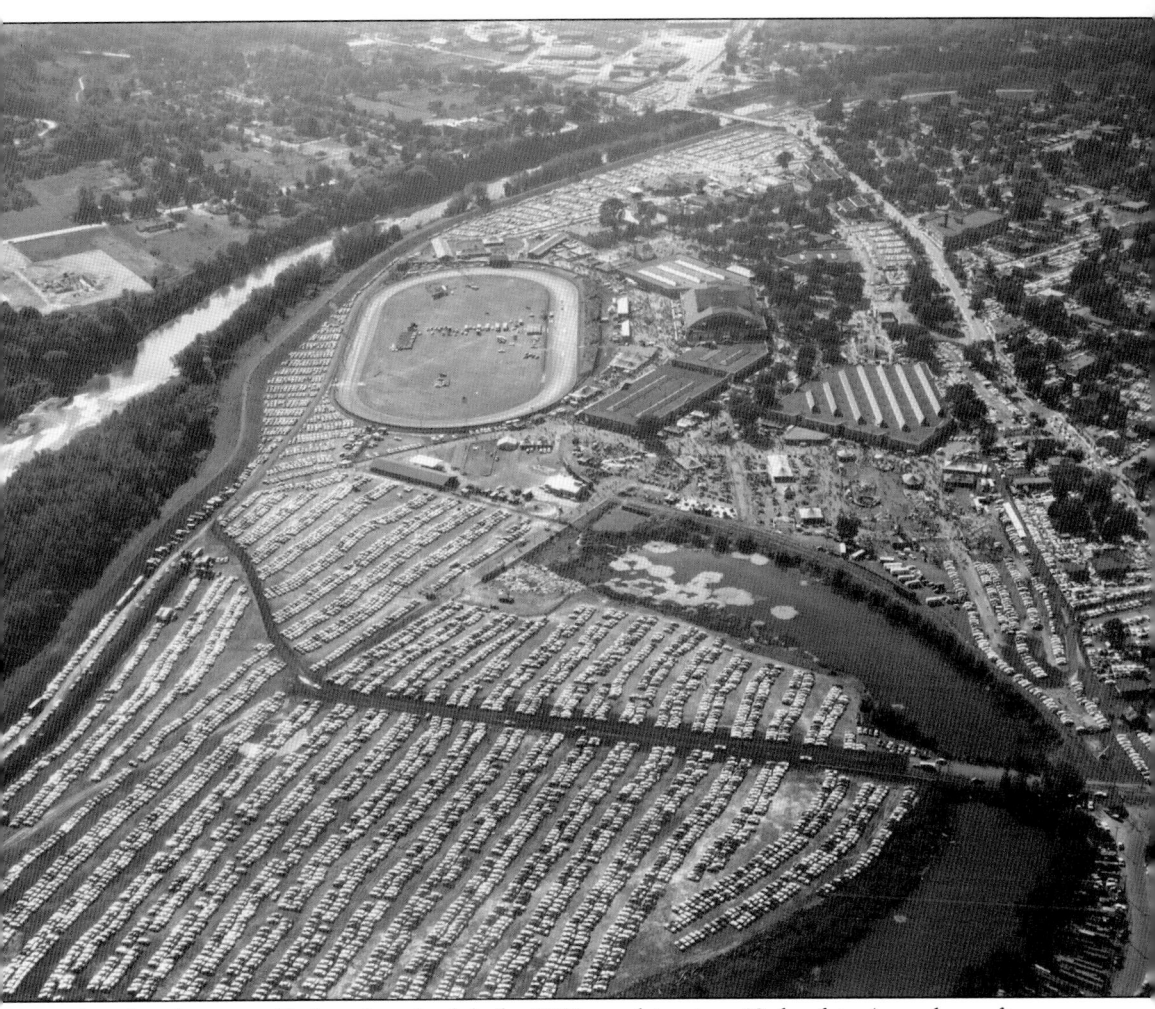

Another day was added to the schedule for 1968, resulting in a 10-day fair. Attendance figures that year included six one-day attendance records (first Saturday–97,221, first Sunday–133,477, Monday–46,004, Wednesday–61,648, second Saturday–115,811, and second Sunday–69,404), record attendance for the final day of the fair (second Sunday), and a total attendance record of 703,034, the largest to date. Helping to draw fairgoers that first Sunday was a Country Western Roundup featuring Buck Owens, Faron Young, Porter Wagoner, Dolly Parton, Waylon Jennings, and Penny Starr. On the racetrack, seen at center, the 250-lap, 125-mile Eastern States Championship for modified stock cars was held on September 14 and won by 1967 Riverside Park Rookie of the Year Ernie Caruso of Bloomfield, Connecticut. On September 21, Roy Halquist of Fairfield, Connecticut, won the 250-lap, 125-mile Grand American Championship for late model stock cars for the second consecutive year. The year 1968 would turn out to be the final one for auto racing at the exposition.

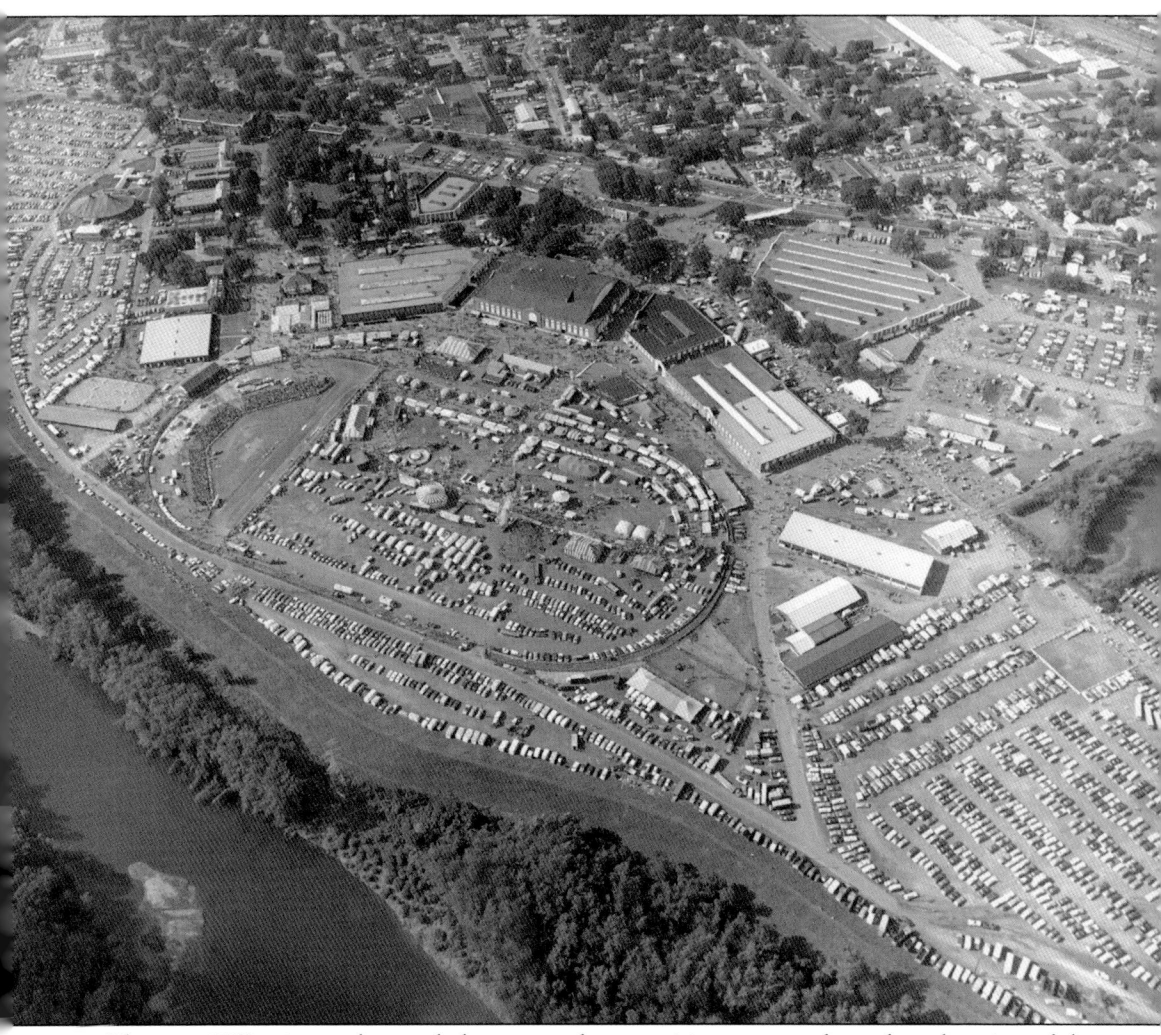

The year 1969 witnessed several changes at the exposition, among them the relocation of the amusement section from its former location near the Better Living Center, seen at top right, to the site of the former racetrack. The new midway was renamed Funland. The eastern end of the old racetrack was kept intact to accommodate the automobile thrill shows and is seen at center left. Referred to as the Outdoor Arena, it was also the location of Gene Holter's Jungleland Animals Show; the Eastern States Lumberjack Championship; and Shindig, the "rocking, rolling concert that features something for everyone," and showcased "the best amateur bands from all over New England." Professional entertainers at the 1969 fair included Anita Bryant, the Cowsills ("The Rain, the Park and Other Things," "Hair"), Merle Haggard, Bobby Vinton, Tahuna and his Polynesian Dancers, and Lorne Greene of *Bonanza* fame. Frances Mitchell, Miss Wool of America, was among the 716,223 people who visited the fair in 1969. That record-breaking figure represented an increase of 13,189 people over the previous year's record total.

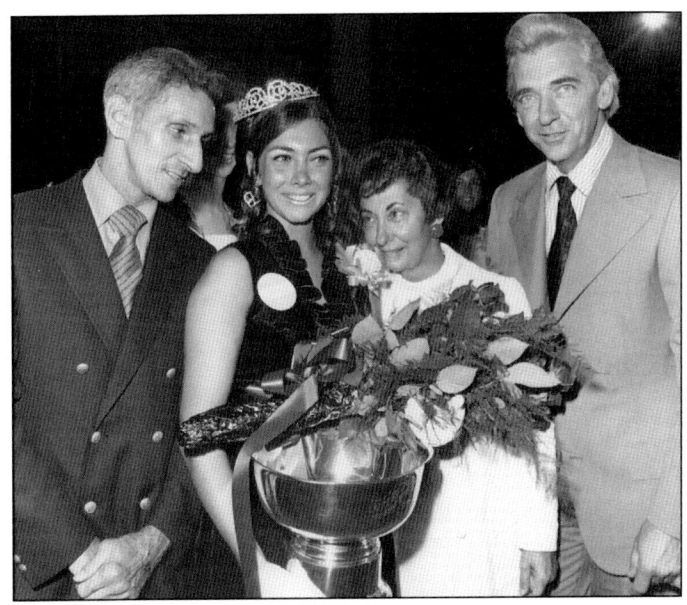

A Festival of College Queens was held by the exposition from 1961 to 1970, with a winner selected from six young ladies representing New England colleges based on academic and personal achievement, career motivation, and health and beauty. Seen with her parents is the 1970 College Queen, University of Maine senior Nancy Pedrini, second from left, and Country Western star Carl Smith, right, who presented her with a tiara and $1,000 scholarship from the exposition.

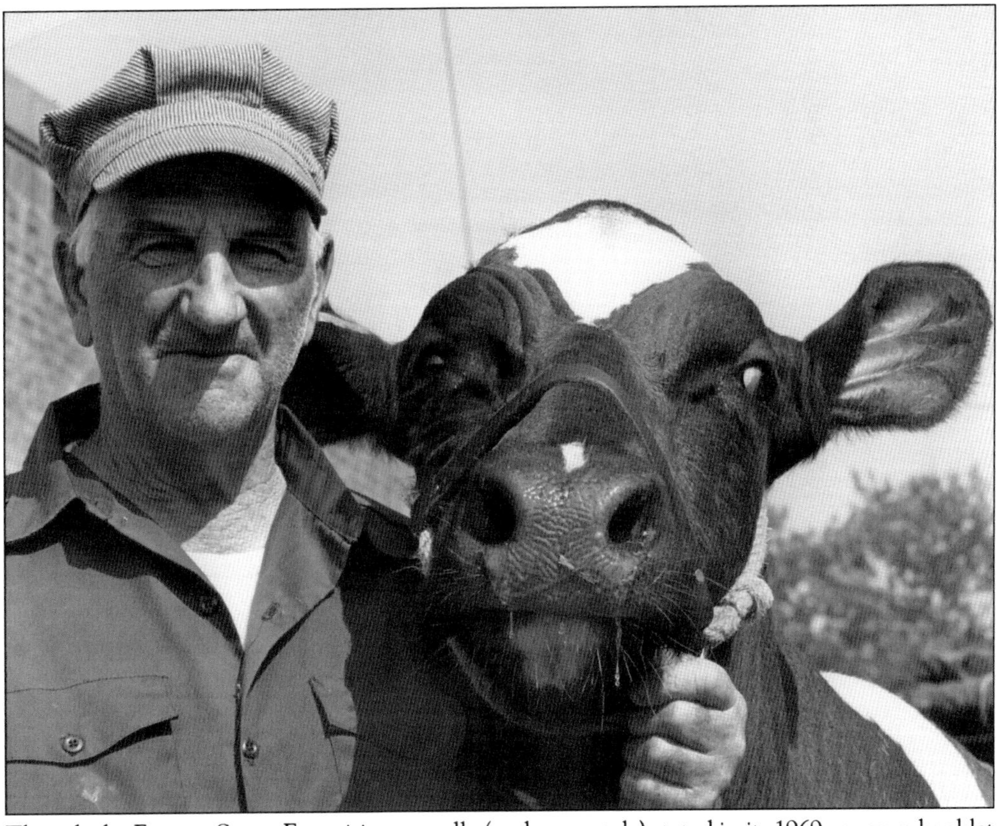

Though the Eastern States Exposition proudly (and accurately) stated in its 1969 program booklet that it "has always placed its primary emphasis on young people, their education and entertainment," the fair welcomes visitors and exhibitors of all ages. Seen here in 1970 is Leon Martel, 84, of Rouse's Point, New York, first-time and oldest exhibitor at the exposition that year.

Massachusetts transportation secretary Alan Altshuler makes a presentation regarding high-speed passenger rail service at the exposition's New England Governors Conference on September 18, 1971, after which the governors voted to seek federal funds for regional high-speed rail service. Forty-five years later, the February 25, 2016, headline of *The Republican* read, "High-speed rail line pitched. Lawmakers considering study of Boston-Springfield route." One cannot make this stuff up.

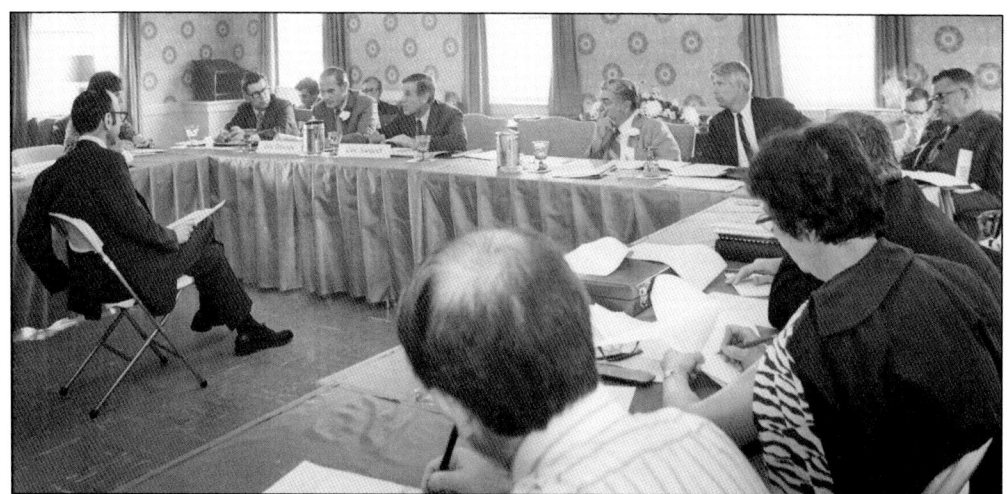

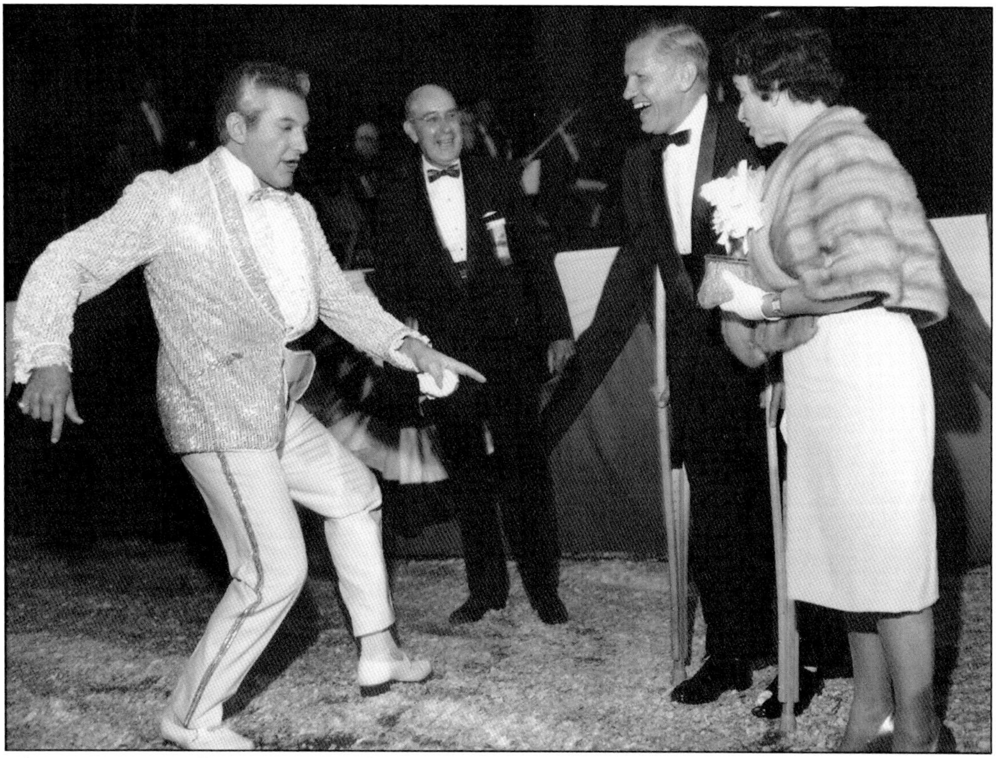

Liberace was a popular attraction at the Storrowton Theater—the Liberace Show appeared annually from 1970 to 1975 and was seen by thousands of area residents, including Agawam's Patty Souder, who met the showman himself and still vividly remembers the encounter. The pianist is seen here demonstrating the Twist to Massachusetts governor John Volpe and his wife, Jennie, at the 1962 Governor's Command Performance while exposition president J. Loring Brooks Jr. looks on.

Doan Helicopter Service of Daytona Beach, Florida, offered helicopter rides at the exposition for several years in Bell 47 helicopters. Current exposition president Gene Cassidy recalled a memorable birthday flight in 1969 with his dad that took him over his grandfather's Agawam farm. The author also spent time aloft with his then girlfriend (now wife) in the 1980s, before the flights were discontinued.

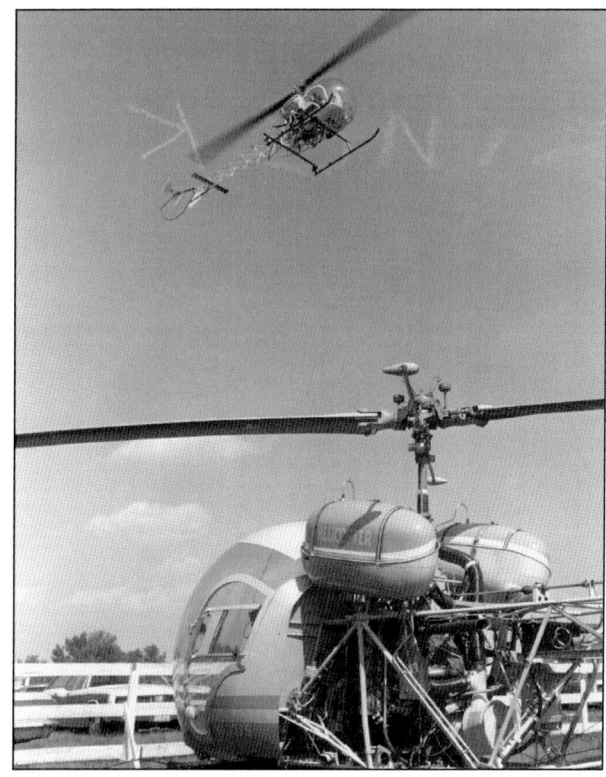

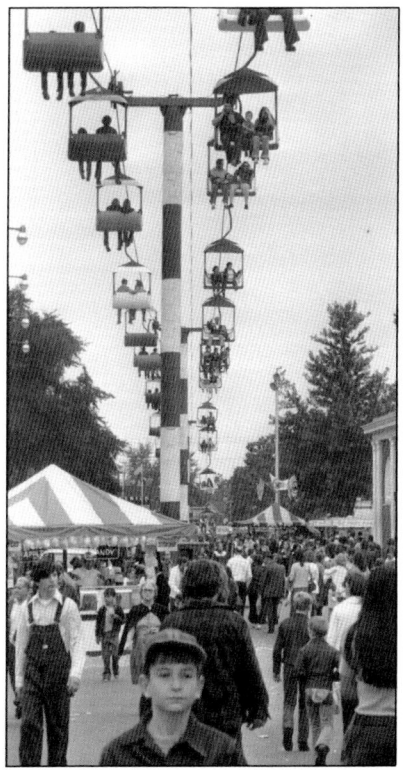

The Sky Lark ride, seen here in 1972, was one of the new attractions that debuted in 1969. The 930-foot chairlift transported visitors between Funland and the area between the Better Living Center and the Brooks Building. The lift was over 40 feet high and held a maximum of 750 people.

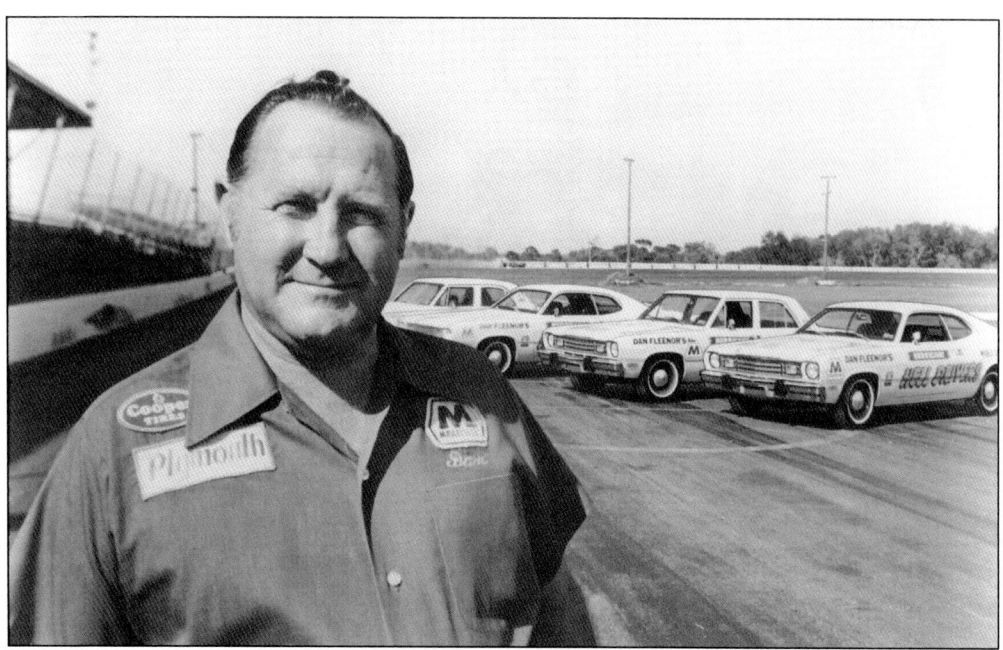

Dan Fleenor and the Hurricane Hell Drivers Auto Stunt Show were the featured automobile thrill show at the fair from 1966 to 1979. Stunts such as the Motorcycle Leap of Death over a row of parked cars; the Slide for Life, where one of the drivers would drop from the rear of a speeding car; and the 60-foot ramp-to-ramp auto jump ensured its popularity with exposition spectators.

Performing at the 1974 exposition were some of the biggest names of the time: Mac Davis, Donna Fargo, Tony Orlando and Dawn, Kenny Rogers and the First Edition, and "the youngest singing star of the Grand Ole Opry," Barbara Mandrell. Described at the time by the Sunday Republican as simply a "blonde bombshell," Mandrell would go on to become one of the most successful female vocalists of the 1970s and 1980s. (Author's collection.)

Anna and Peter Ortolani, proprietors of the Yankee Boy food concessions at the exposition, created the Ortolani Scholarship in 1968. The first $1,000 scholarship was awarded to West Springfield High School graduate Robert R. Fish for outstanding academic achievement and leadership. The Ortolanis are seen here in 1976 celebrating their 50th year with the exposition.

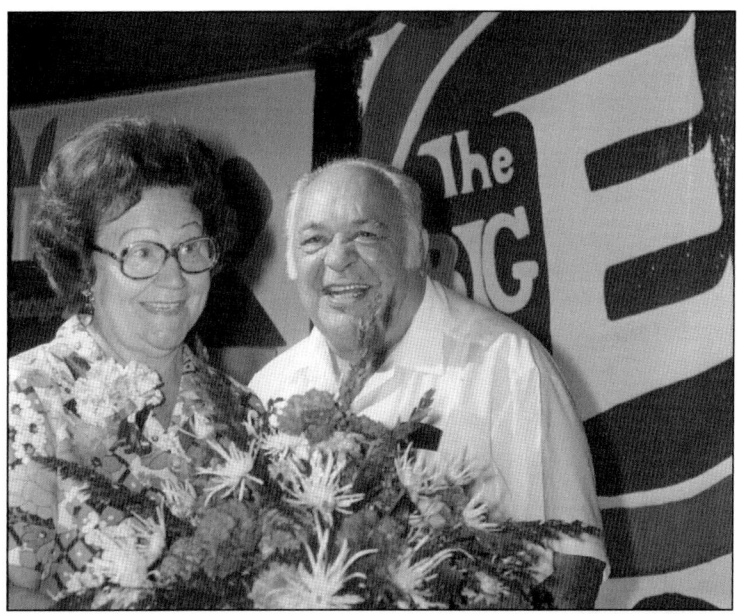

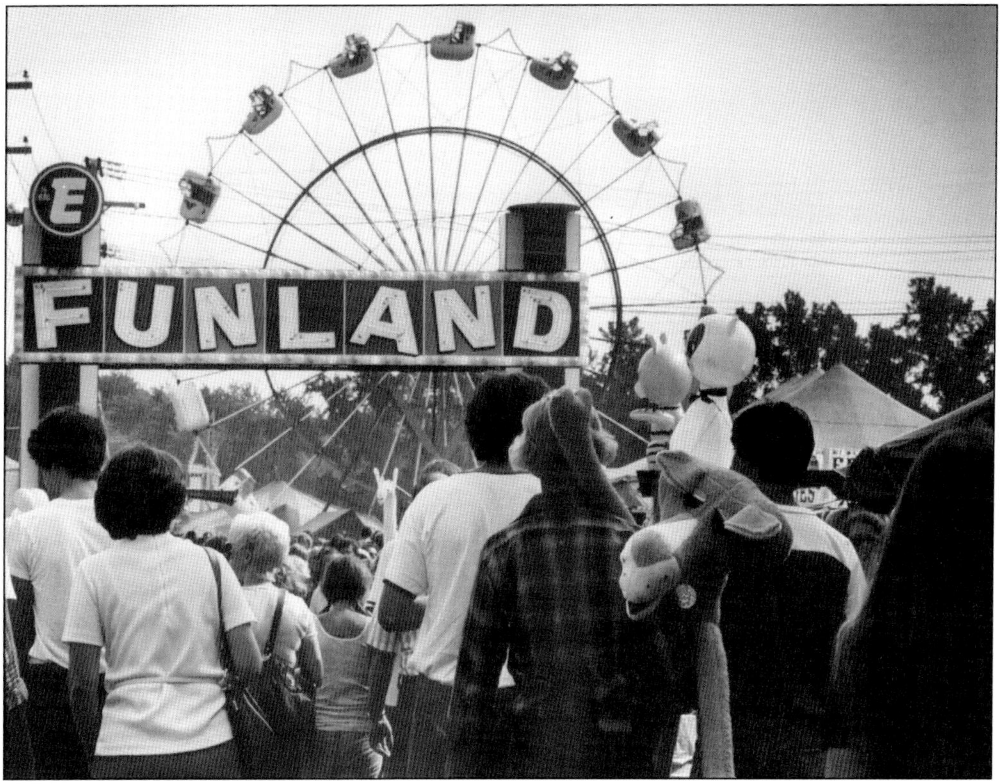

In the late 1970s, Funland featured more than 40 rides and shows, including several Ferris wheels, the Yo-Yo, the Zipper, and the classic Musik Express, making it "probably the largest midway that's ever been in the New England area," according to Dennis G. Kopcha, co-owner of East Coast Shows of Pennsylvania, who operated the midway at the time. Beginning in 1975, fairgoers had 12 full days to enjoy the fair.

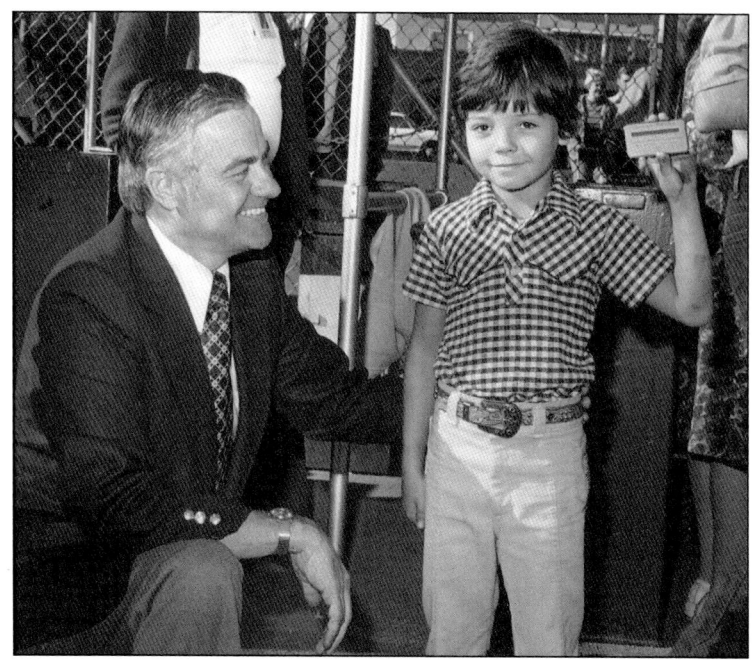

Attendance at the exposition increased steadily throughout the 1970s, from 700,000 visitors in 1973 to 929,000 in 1976. Here, seven-year-old Chad Yergeau of Monson, Massachusetts, displays the gold lifetime pass presented to him by Eastern States Exposition general manager George W. Jones, left, for being the one-millionth visitor to the 1978 fair. Total attendance that year was 1,006,863.

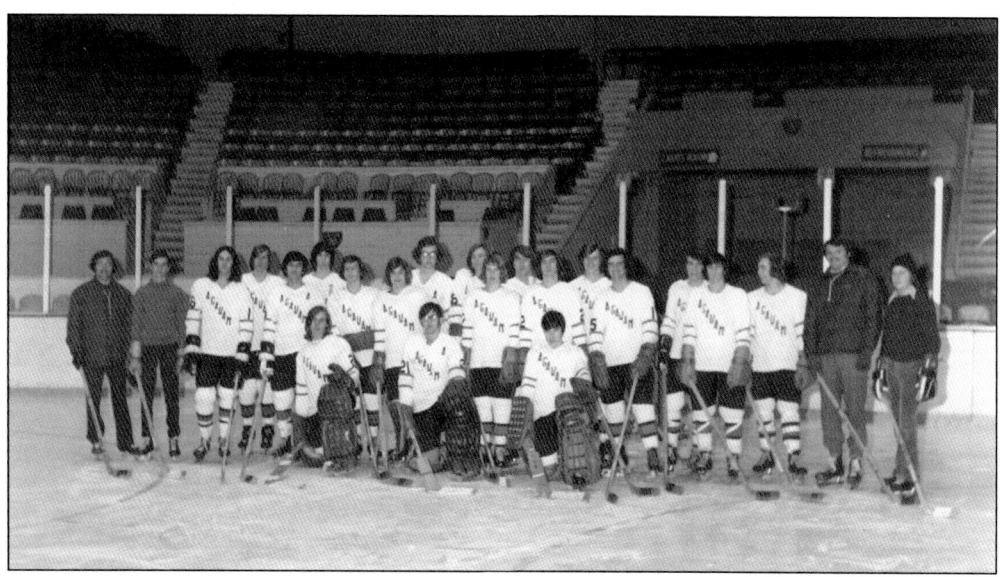

The 1972–1973 Agawam High School hockey team is pictured on the Coliseum ice. From left to right are (kneeling) Gary Brown, Doug Janik, and Brian Colby; (standing) coach Russ Ramah, manager Al Sapelli, Jamie Fenton, Mike Kerr, Bud Ramah, Brian Keeley, Jeff Nolin, Steve Gould, Brian Telford, Tom Fenton, Randy Economidy, John Marino, Rick Loncrini, Scott Brown, Barry Economidy, Tom Ennis, Tom Dalmolin, Dave Mason, assistant coach Art Gage, and manager Tony Veteramo. (Courtesy of Ellen Janik.)

Hundreds (maybe thousands) of musical groups of every stripe and description have entertained fairgoers at The Big E since Short's Concert Band opened the National Dairy Show on Thursday, October 12, 1916, with its rendition of the "Eastern States Exposition March." Members of the Riverdale Gardenaires Nursing Home Kitchen Band are seen here in the early 1980s on the Storrowton Village gazebo stage performing with a variety of improvised and homemade instruments.

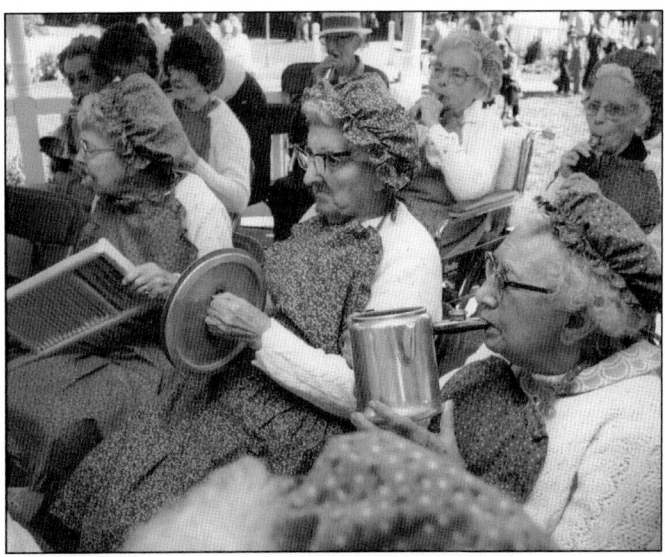

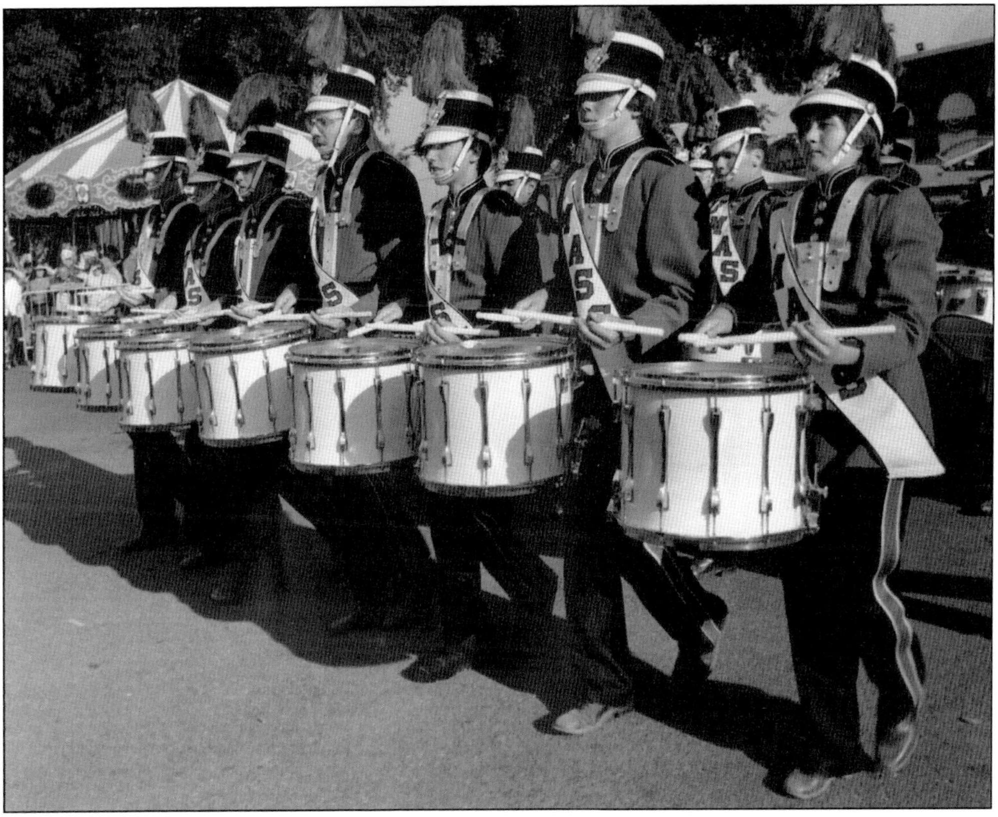

"Mass. Aggie" and the exposition have deep roots—Kenyon L. Butterfield, president of Massachusetts Agricultural College (now the University of Massachusetts Amherst), was a supporter from the start. Here, the University of Massachusetts Minuteman Marching Band drumline leads the band in a 1984 parade. A 2016 appearance by the 400-member University of Massachusetts Marching Band would not only be a fitting reconnection of institutions, but a true thrill for anyone lucky enough to catch their performance.

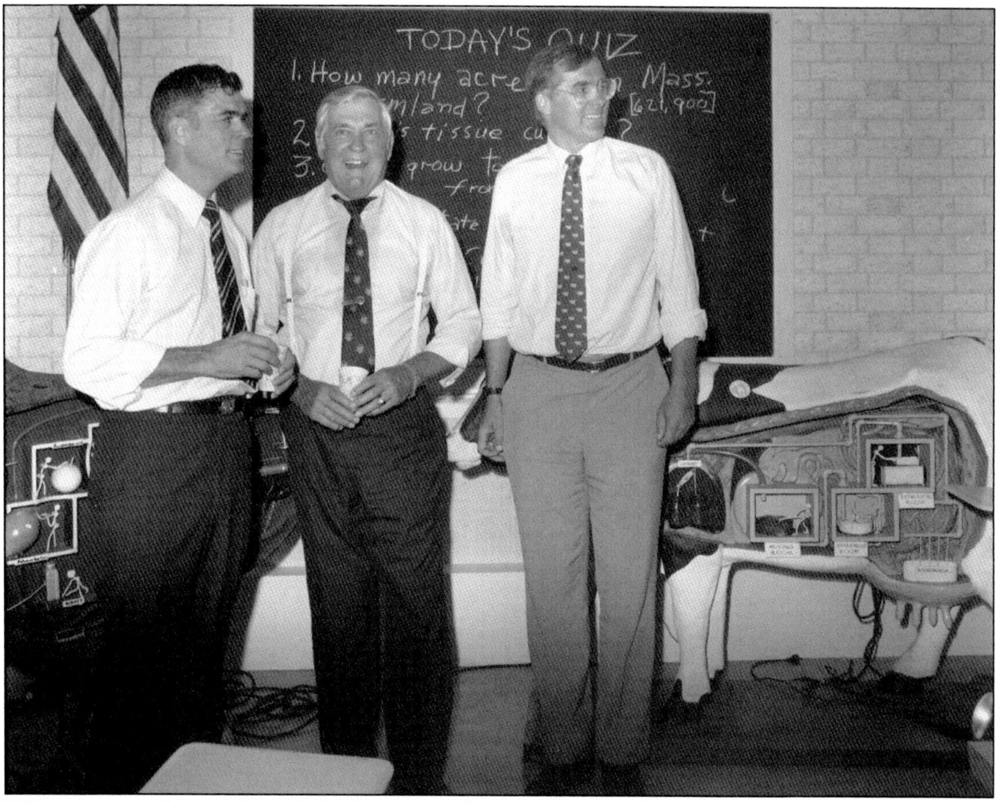

US secretary of agriculture John R. Block, left, and Massachusetts first district congressman Silvio O. Conte, center, were given a tour of the Massachusetts Building's agricultural exhibits on September 20, 1985, by August Schumacher, Massachusetts commissioner of food and agriculture. Conte felt the mid-1980s were "one of the worst times ever" for farmers, citing excessive supply, low prices, and shrinking land values.

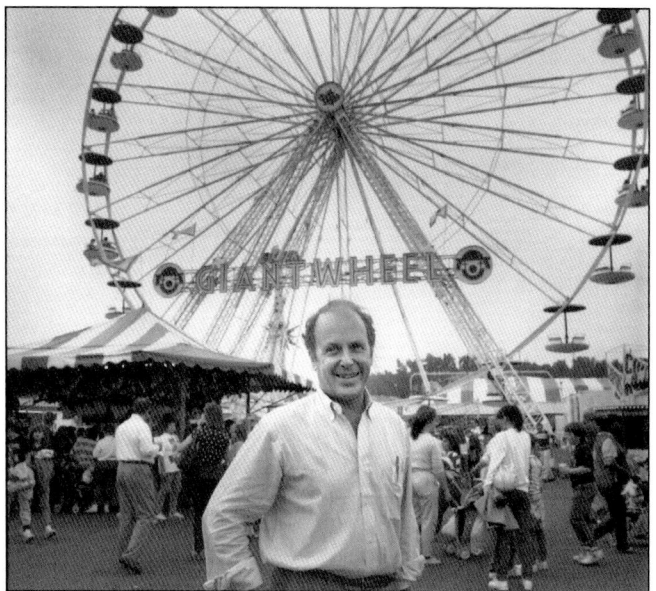

Conklin Shows, the largest traveling midway in North America at the time, operated its Magic Midway at The Big E from 1979 to 2004. Seen here in 1988 is James Conklin in front of the popular Giant Wheel—the largest portable Ferris wheel in North America. Since 2005, the midway has been operated by the world's largest traveling outdoor amusement park, North American Midway Entertainment of Farmland, Indiana.

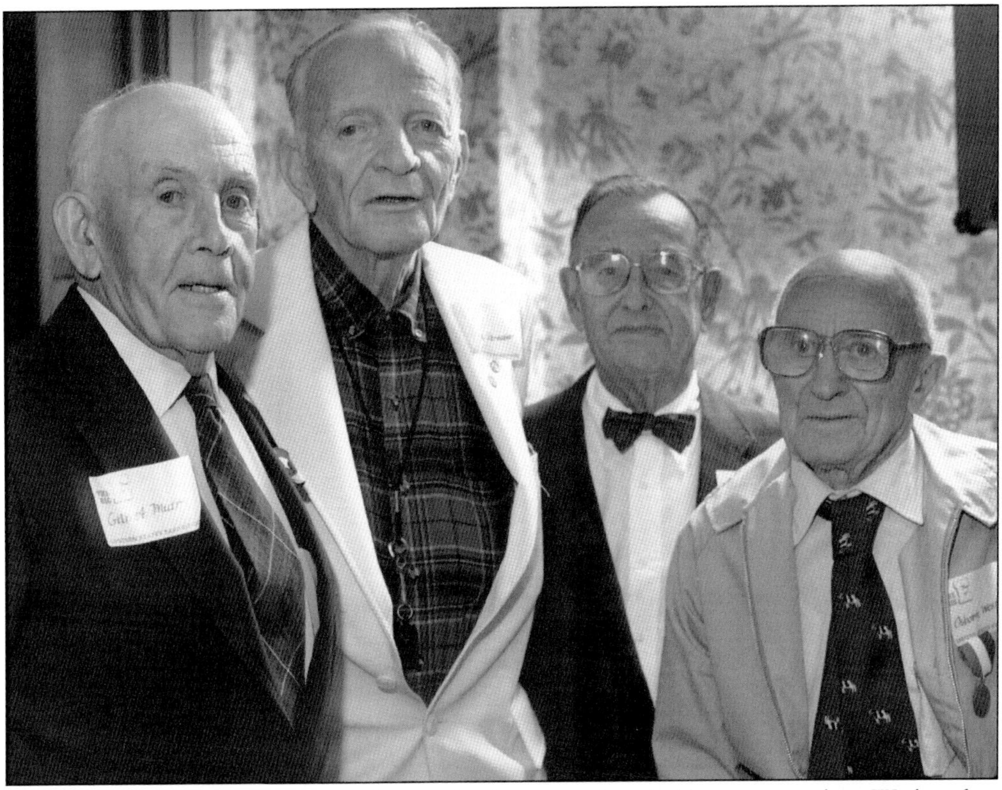

To commemorate the epic event that took place exactly a half century earlier, Wednesday, September 21, 1998, was designated Hurricane of '38 Day. Among those at the fair when the storm hit were, from left to right, Gilbert Muir of Hope, Rhode Island; Richard Streeter of Longmeadow, Massachusetts; Dr. Hilton Boynton of Richmond, Massachusetts; and Osborne West of Hadley, Massachusetts, all of whom gathered at Storrowton Tavern for a reunion.

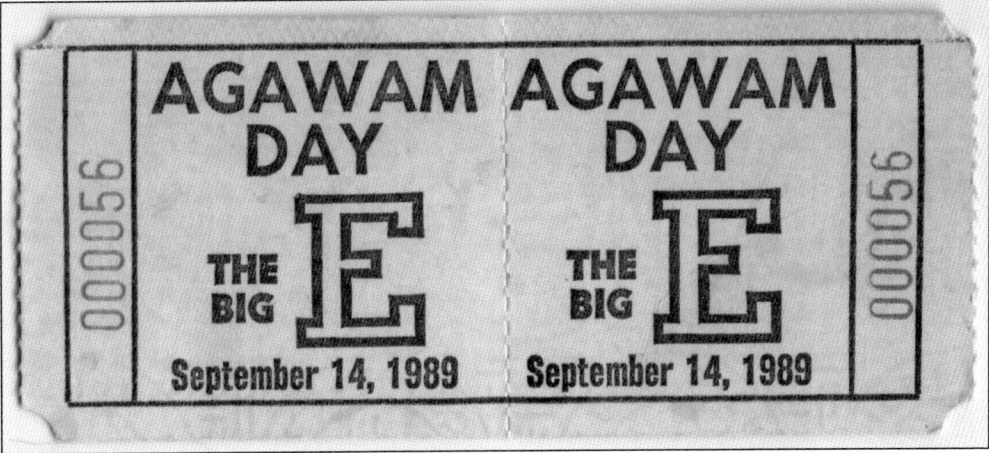

The first Agawam Day was held by the exposition on September 14, 1989, to honor its neighbor across the river. The Agawam Congregational Church and the Agawam Lions Club were both honored with Friend of The Big E awards for decades of providing meals to fairgoers. The Lions Club continues to offer its famous barbecue chicken dinner; their new "conewich" became the most popular food item at the 2015 fair. (Author's collection.)

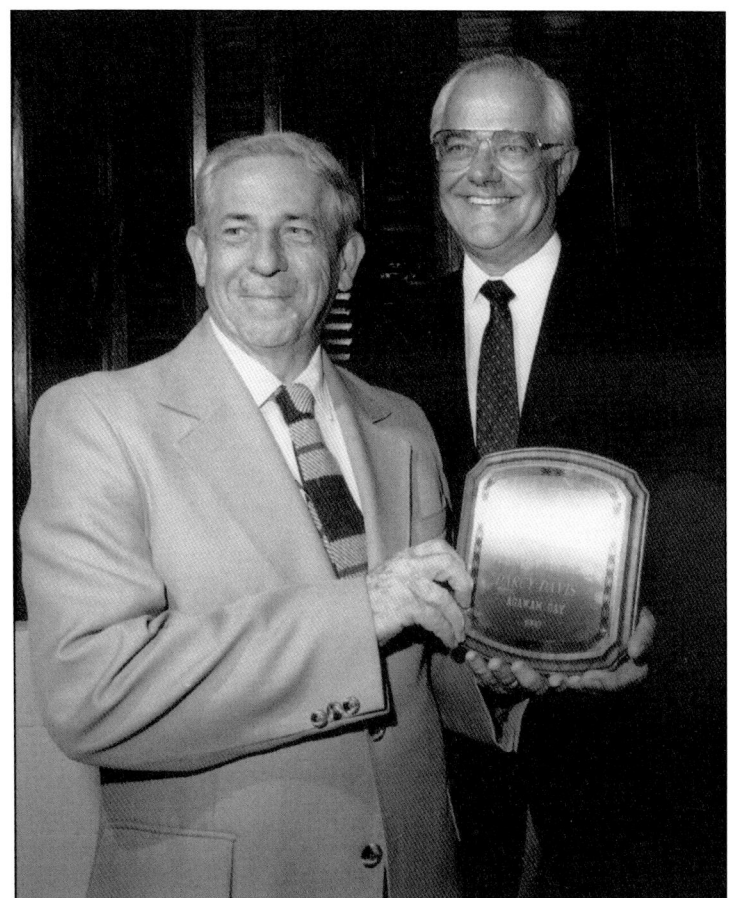

Eastern States Exposition president George Jones presents Darcy B. Davis Jr. with the 1990 Friend of The Big E award. Jones would announce his retirement in 1990 after 21 years with the fair, 14 as general manager/president. Jones had first visited the fair in 1948 as a 16-year-old 4-H exhibitor. Davis was the longtime Agawam High School band director and would retire that year after 35 years as "Agawam's Music Man."

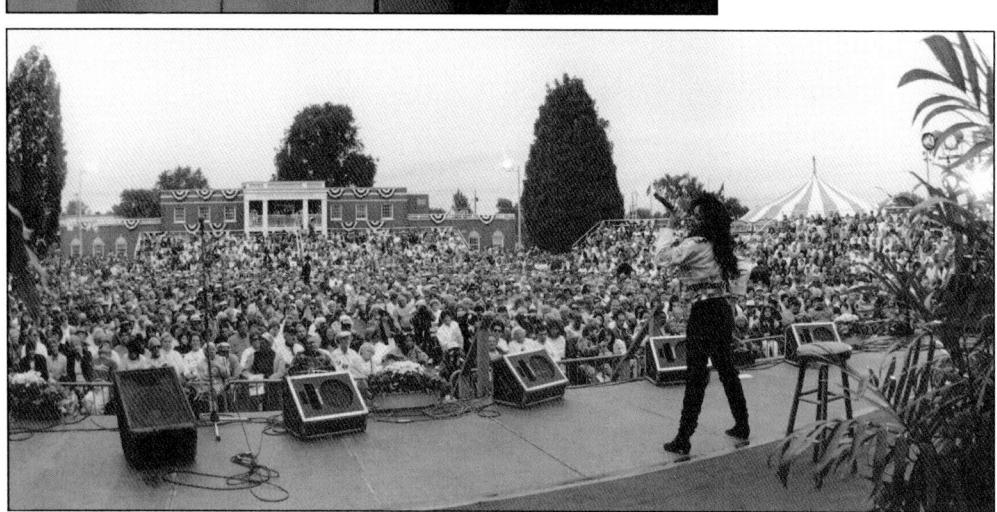

La Toya Jackson interrupted her 1990 world tour to perform on The Big E's Miller Genuine Draft Bandshell stage, flying in from Germany to perform four shows on September 19 and 20. She flew to London the following day to resume her tour. Also appearing at the fair that year were Tammy Wynette, Charley Pride, and Lorrie Morgan. BiggiE, the fair mascot, celebrated his 10th birthday in 1990 as well.

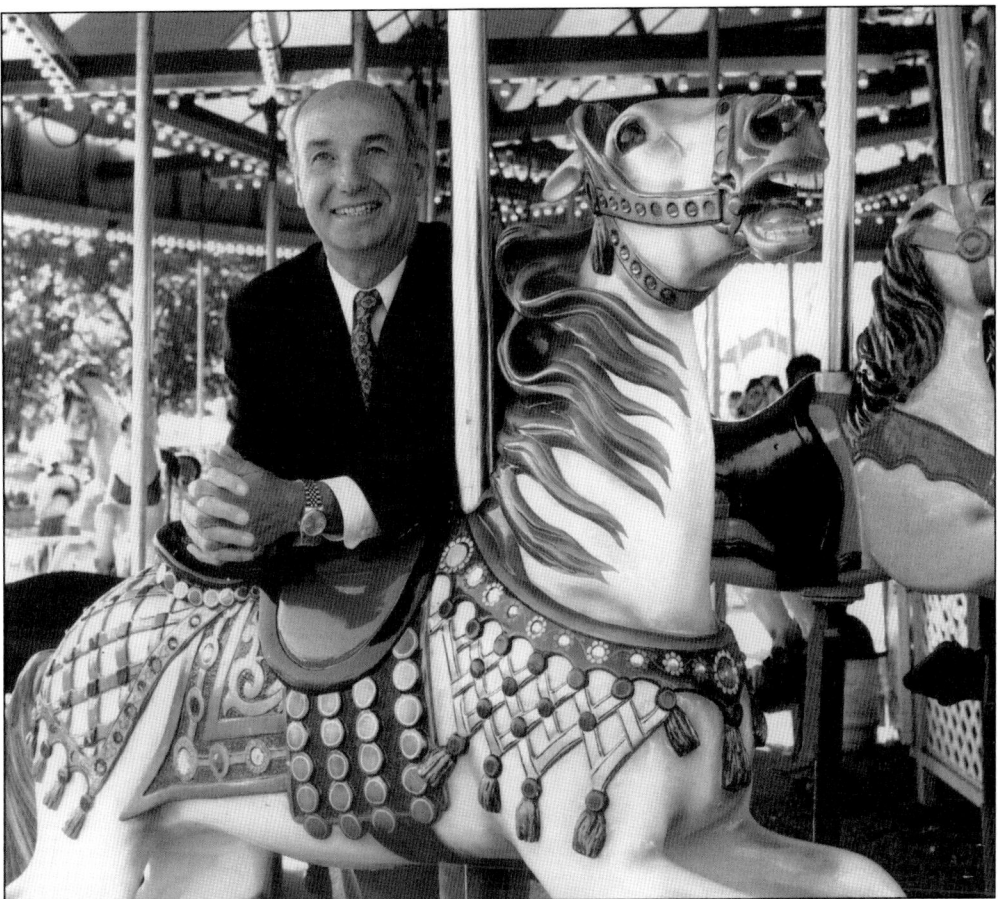

Wayne McCary took the reins of the exposition in 1991 and served as the organization's president until 2012. He had served as executive assistant at the fair from 1973 to 1983 and executive vice president from 1986 to 1990. Under his leadership, more than $36 million in capital improvements were made to the exposition grounds and more than 40 million fairgoers visited the annual fall fair. At the time of McCary's retirement, more than 120 shows and events were held on the grounds throughout the year, and the fair saw annual attendance of approximately 1.2 million people. In 1994, the fair was extended to its current 17 days, and McCary is credited with bringing the popular Mardi Gras parade to the fair in 2000. McCary is also responsible for the creation of The Big E Super Circus. He served as president of the Massachusetts Agricultural Fairs Association in 1993 and was named to its Hall of Fame in 2010. He also served as president of the International Association of Fairs and Expositions in 1997 and was named to its Hall of Fame in 2001.

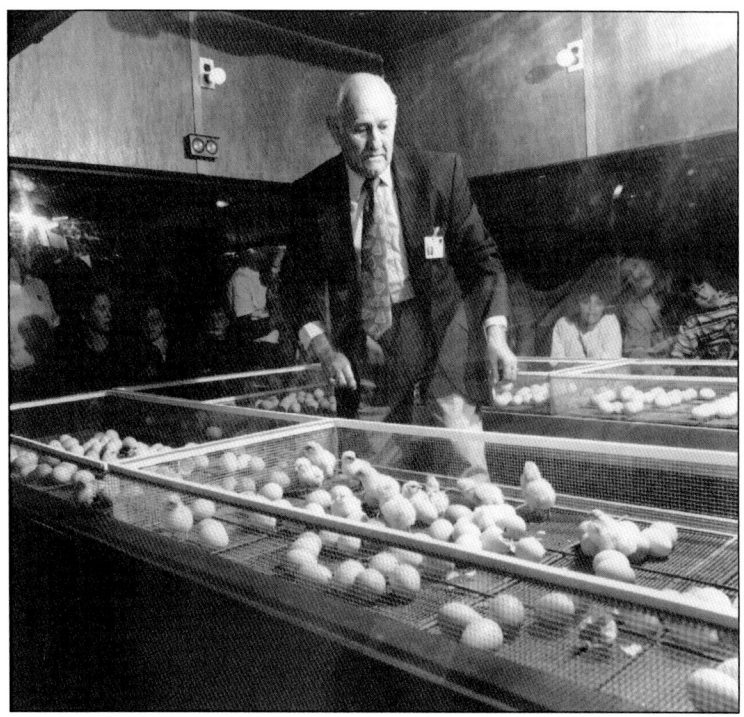

In 1995, the late Harry Adriance of Amherst, Massachusetts, is pictured inside the Farm-A-Rama chick hatchery that he developed in the 1970s. Over the run of the fair, thousands of chicks would be hatched, to the delight of fairgoers of all ages. In 2015, an interactive virtual poultry display replaced the chicks due to concerns over the spread of avian flu. Fair officials expect that live chicks will return for the fair's centennial edition.

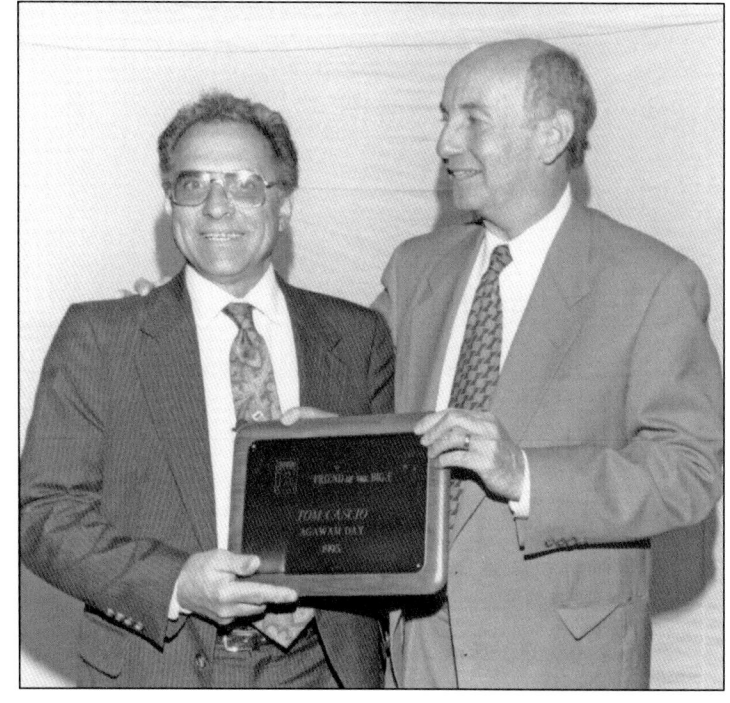

Eastern States Exposition president Wayne McCary presents the 1995 Friend of The Big E award to Thomas Cascio of Agawam. Other recipients of the award over the years have included Agawam school superintendent James V. Bruno, longtime dairy barn volunteer Zoafia Demko, and Jim Fenton, for his assistance in planning the American Truck Historical Society's National Convention and Antique Truck Show that was held on the exposition grounds in 2012.

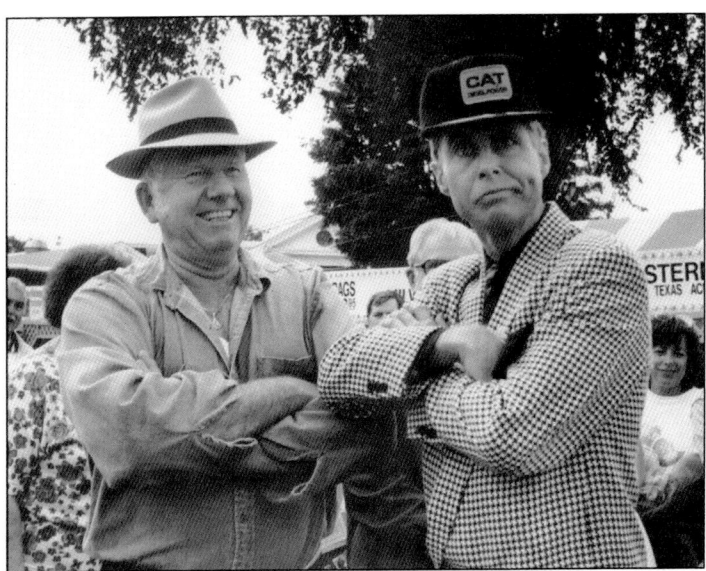

Douglas Sherman, left, of Springfield, and the "Mechanical Man," Randy Burns, swap hats during the 1995 fair. The self-proclaimed "best robot impersonator in the world" has been one of the favorite performers at the fair for more than two decades. Burns's 10-minute act consists of "robotic moves and comedic timing" as he steals hats, kisses women's hands, and teases random fairgoers.

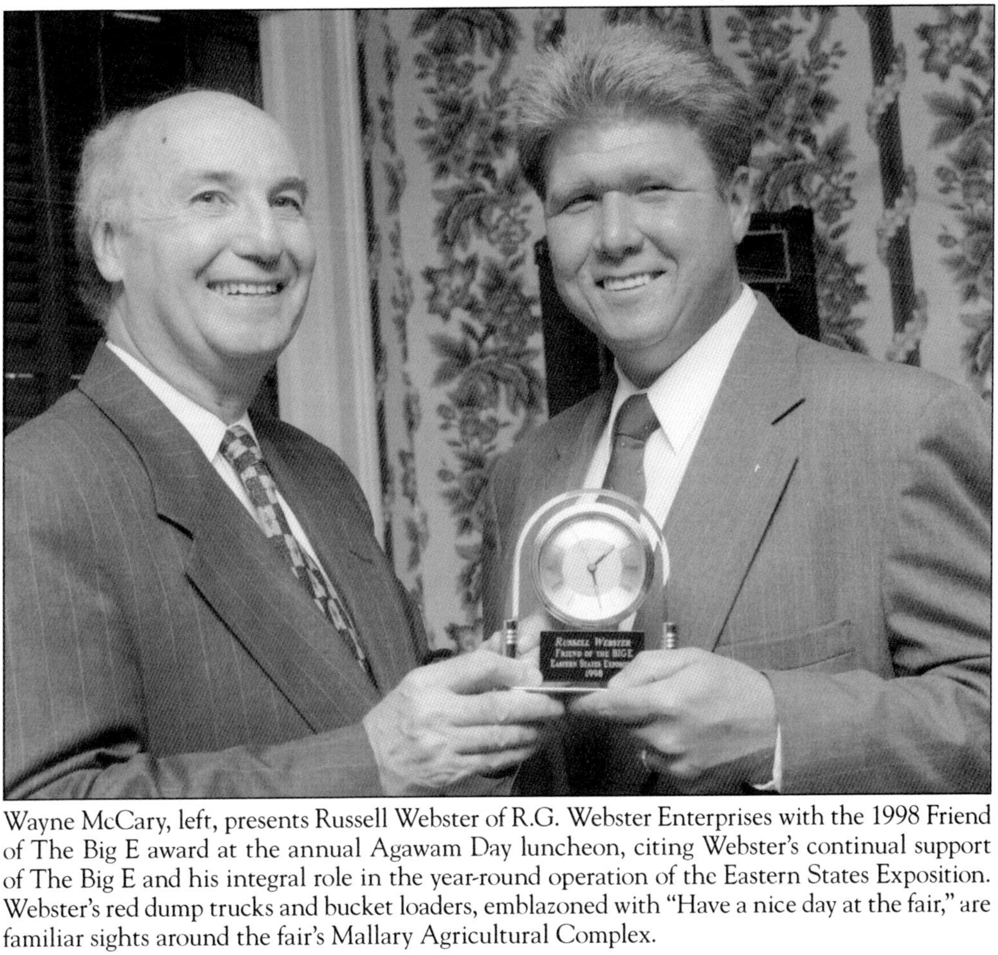

Wayne McCary, left, presents Russell Webster of R.G. Webster Enterprises with the 1998 Friend of The Big E award at the annual Agawam Day luncheon, citing Webster's continual support of The Big E and his integral role in the year-round operation of the Eastern States Exposition. Webster's red dump trucks and bucket loaders, emblazoned with "Have a nice day at the fair," are familiar sights around the fair's Mallary Agricultural Complex.

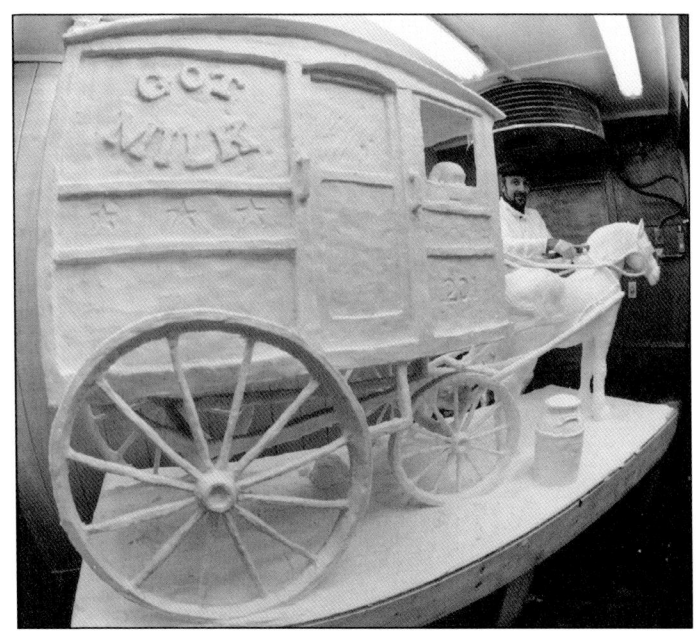

A graduate of the sculpture department of the Pennsylvania Academy of the Fine Arts, Jim Victor of Conshohocken, Pennsylvania, is seen in the Mallary Agricultural Complex's refrigerated butter barn crafting a milk wagon out of butter during the 1998 Big E. Victor and his wife, Marie Pelton, both well-known food sculptors, also work in more traditional materials such as bronze, steel, and cast stone.

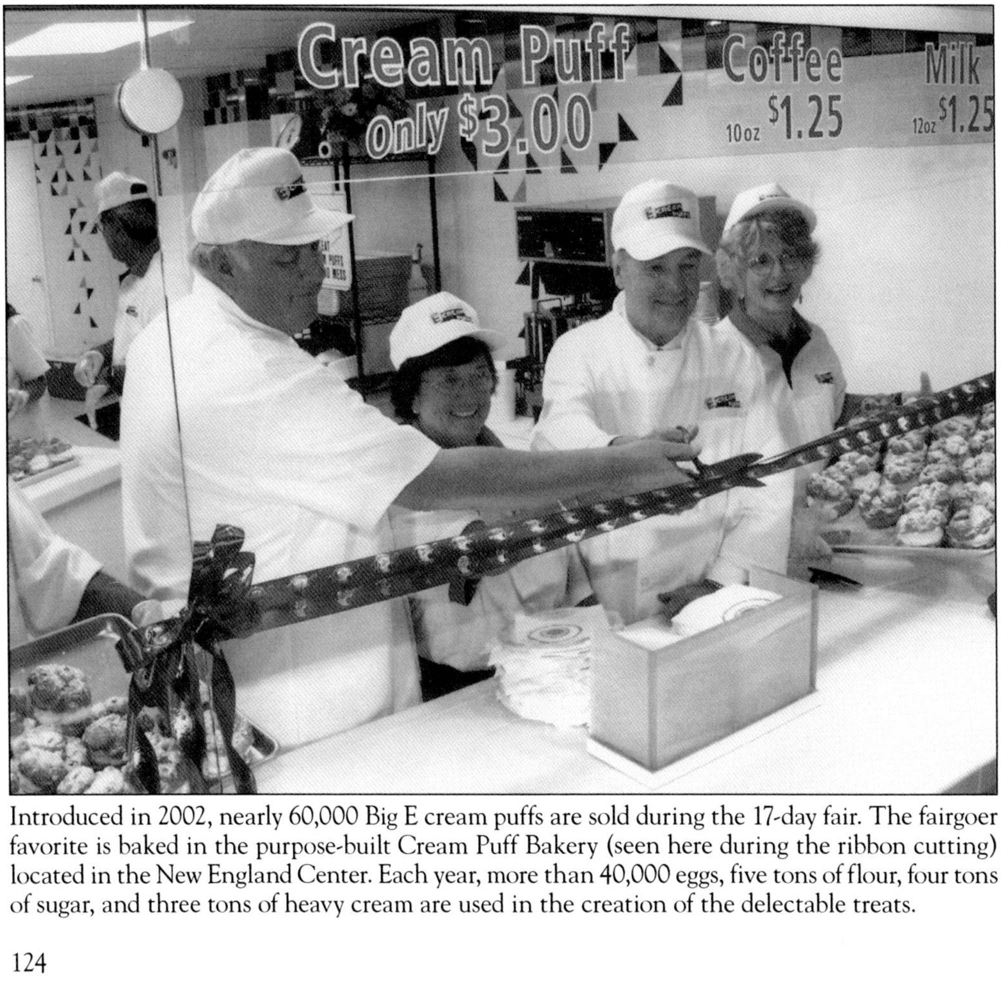

Introduced in 2002, nearly 60,000 Big E cream puffs are sold during the 17-day fair. The fairgoer favorite is baked in the purpose-built Cream Puff Bakery (seen here during the ribbon cutting) located in the New England Center. Each year, more than 40,000 eggs, five tons of flour, four tons of sugar, and three tons of heavy cream are used in the creation of the delectable treats.

The Cecchi family has been farming in Agawam and Feeding Hills, Massachusetts, since the 1920s. The fifth generation was well represented at The Big E's 2003 Youth Giant Pumpkin and Squash Contest. From left to right are Megan Cecchi, Joseph Cecchi, Jordan Venne, Bailey Cecchi, and Dillon Cecchi. Joseph is currently pursuing a bachelor's degree in sustainable food and farming at the University of Massachusetts Amherst's Stockbridge School of Agriculture. (Author's collection.)

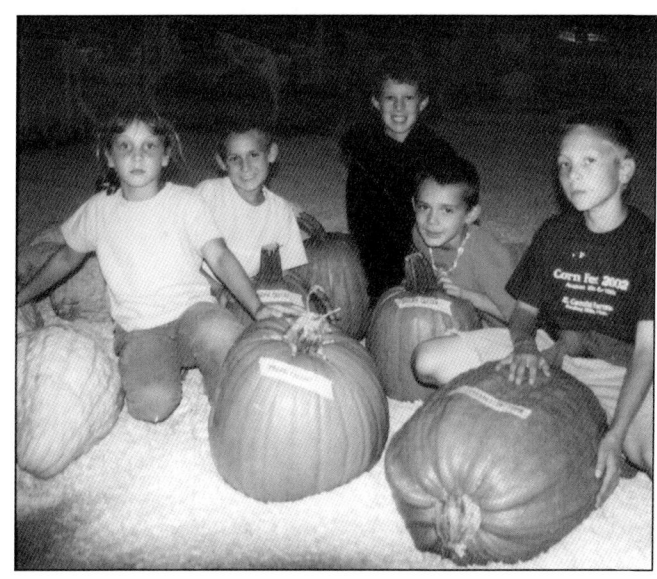

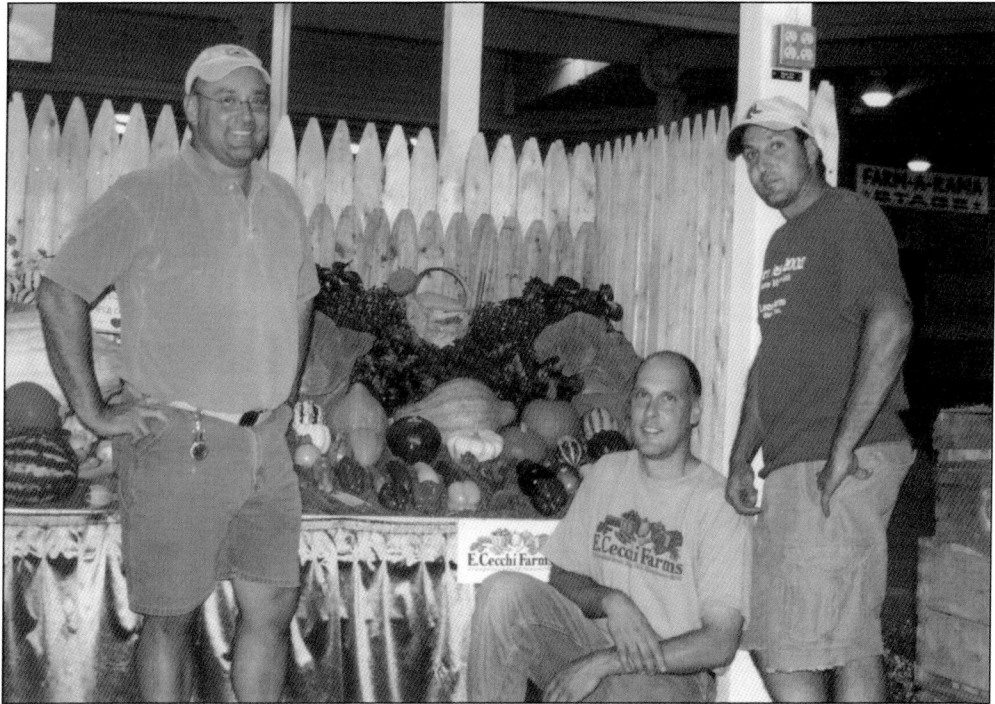

E. Cecchi Farms has exhibited vegetables grown on their Feeding Hills farm nearly continuously since the early 1970s, when the Native Produce Display was located in the long-gone Cuckler Building. The author, at left, and his brothers Bobby (center) and Michael, seen here in 2005, have carried on the tradition and are justifiably proud of the collection of blue ribbons they have earned over the years. (Author's collection.)

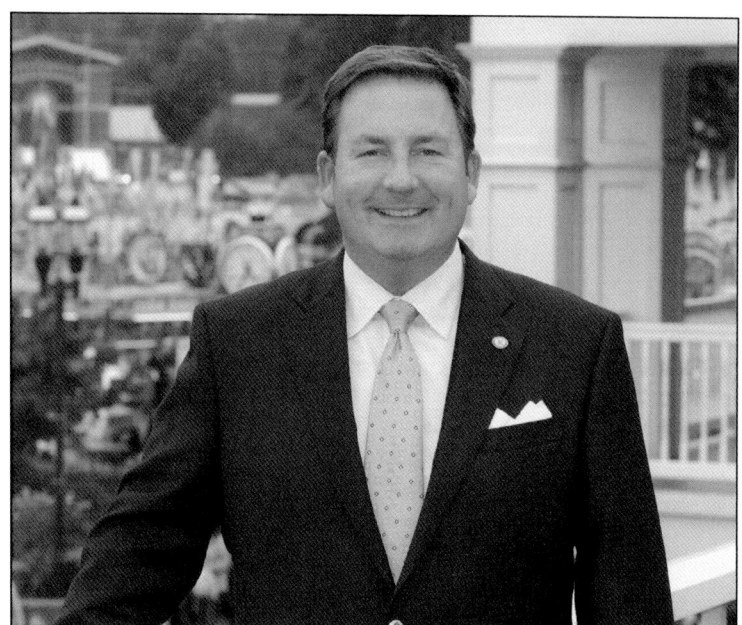

Eugene J. Cassidy grew up less than two miles from the exposition fairgrounds. He joined the organization as director of finance in 1993 and was named its seventh president in 2012. Cassidy's love of the exposition and his deep appreciation of its history and mission are evident. The exposition is in good hands as it begins its second century.

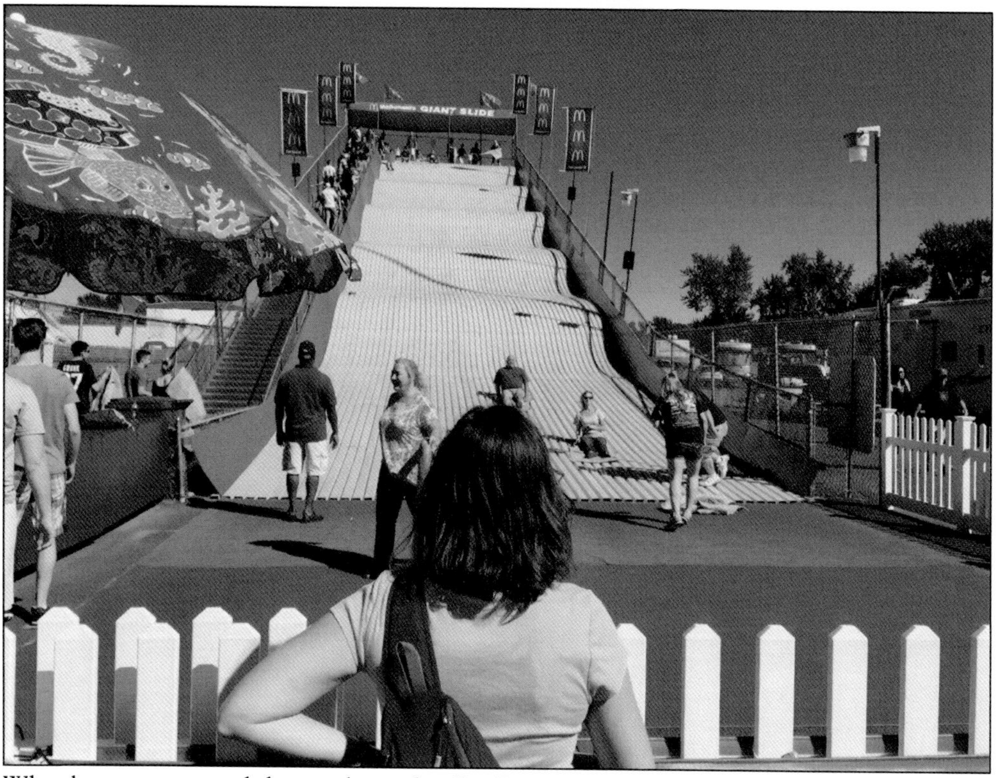

What better way to end than with another Big E icon? In 1968, the first big slide was the blue, four-story Sky-Slide. The current slide was installed near Gate 5 in 1969, relocated east of the Better Living Center in 1983, and moved again to its present location in 1996. Now known as the McDonald's Giant Slide, the 52-foot-tall yellow slide has been a favorite of generations of fairgoers. (Courtesy of Laurie Cecchi.)

Bibliography

Anniversary Book. Springfield, MA: Springfield, Massachusetts Three Hundredth Anniversary Committee, 1936.

Brooks, J. Loring Jr. *"J.L." A Biography*. Springfield, MA: Brooks Bank Note Company, 1952.

Ernest, Thomas K. "The Future of the Eastern States (1.)." *Journal of Education*. December 11, 1919: 601.

Fair Programs, Brochures, misc. Ephemera, Eastern States Exposition Archives, West Springfield, MA and collection of David Cecchi, Agawam, MA.

Gagnon, France M. *Eastern States Exposition 1916–1996, An Illustrated History at 75 Years*. West Springfield, MA: Eastern States Exposition, 1996.

Hinshaw, Kenneth. *4-H, A Story–Weaving Together Actual 4-H Experiences, Historical Sketches of Boys' and Girls' Club Work, and Chronicles of Important 4-H Events*. New York, NY: Orange Judd Publishing Company Inc., 1935.

Kinnison, H.B. "The New England Flood of November, 1927." Water Supply Paper 636-C, 1929, US Department of the Interior.

Lane, Harry L., *Achievement is My Goal*. West Springfield, MA: Horace A. Moses Foundation, 1956.

The Springfield Republican Historical Archive, 1844–1989. www.newsbank.com

The Springfield Republican Historical Archive, 1988–2016. www.newslibrary.com

Tercentenary of New England Agriculture 1630–1930. Commissioners of Agriculture of the Six New England States, 1930.

Scheuerle, John A. "The Revitalizing of Rural Life in New England, Parts I, II, III." *Western New England Magazine*. July 1913: 284–290, August 1913: 356–360, September 1913: 399–404.

———. "Hartford, Forward! Its Origin, History, and Purpose." *Vermonter*. January-February 1912: 467–470.

Discover Thousands of Local History Books Featuring Millions of Vintage Images

Arcadia Publishing, the leading local history publisher in the United States, is committed to making history accessible and meaningful through publishing books that celebrate and preserve the heritage of America's people and places.

Find more books like this at
www.arcadiapublishing.com

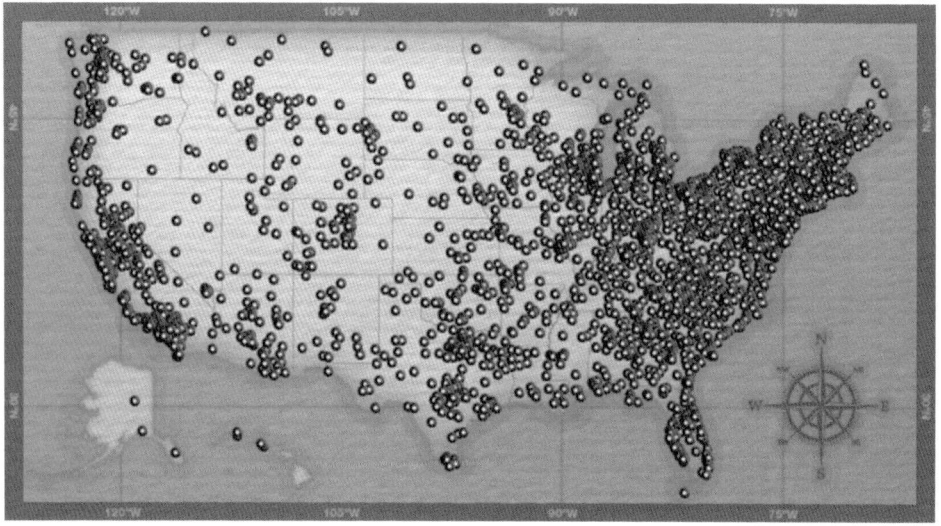

Search for your hometown history, your old stomping grounds, and even your favorite sports team.

Consistent with our mission to preserve history on a local level, this book was printed in South Carolina on American-made paper and manufactured entirely in the United States. Products carrying the accredited Forest Stewardship Council (FSC) label are printed on 100 percent FSC-certified paper.

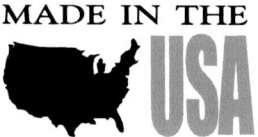